C000137224

RURAL POLICING AND POLICING THE RURAL

Perspectives on Rural Policy and Planning

Series Editors:
Andrew Gilg, University of Exeter and University of Gloucestershire, UK
Henry Buller, University of Exeter, UK
Owen Furuseth, University of North Carolina, USA
Mark Lapping, University of South Maine, USA

Other titles in the series

Rural Policing and Policing the Rural

A Constable Countryside?

Edited by

ROB I. MAWBY
University of Gloucestershire, UK

RICHARD YARWOOD
University of Plymouth, UK

ASHGATE

© Rob I. Mawby and Richard Yarwood 2011

All rights reserved. No part of this publication may be reproduced, stored in a retrieval system or transmitted in any form or by any means, electronic, mechanical, photocopying, recording or otherwise without the prior permission of the publisher.

Rob I. Mawby and Richard Yarwood have asserted their right under the Copyright, Designs and Patents Act, 1988, to be identified as the editors of this work.

Published by
Ashgate Publishing Limited
Wey Court East
Union Road
Farnham
Surrey, GU9 7PT
England

Ashgate Publishing Company
Suite 420
101 Cherry Street
Burlington
VT 05401-4405
USA

www.ashgate.com

British Library Cataloguing in Publication Data
Rural policing and policing the rural : a constable
countryside?. -- (Perspectives on rural policy and
planning)
1. Police, Rural. 2. Police, Rural--Cross-cultural
studies. 3. Rural crimes--Prevention. 4. Rural crimes--
Investigation.
I. Series II. Mawby, R. I. III. Yarwood, Richard.
363.2'091734-dc22

Library of Congress Cataloging-in-Publication Data
Rural policing and policing the rural : a constable countryside? / [edited] by Rob I. Mawby and Richard Yarwood.
    p. cm. -- (Perspectives on rural policy and planning)
Includes bibliographical references and index.
ISBN 978-0-7546-7473-3 (hbk) -- ISBN 978-1-4094-2004-0 (ebook) 1. Police, Rural. 2. Rural crimes. I. Mawby, R. I. II. Yarwood, Richard.
HV7965.R87 2010
363.2'3091734--dc22

2010027859

ISBN 9780754674733 (hbk)
ISBN 9781409420040 (ebk)

Mixed Sources
Product group from well-managed
forests and other controlled sources
www.fsc.org Cert no. SGS-COC-2482
© 1996 Forest Stewardship Council
FSC

Printed and bound in Great Britain by
TJ International Ltd, Padstow, Cornwall

Contents

PART II   POLICING THE RURAL

List of Figures

List of Tables

For Ruth, Elizabeth and William and in loving memory of my mother.
Richard Yarwood

For Maureen.
Rob Mawby

Notes on Contributors

Elaine Barclay is Senior Lecturer in Criminology, University of New England, Armidale NSW. For 15 years, she has undertaken rural social research on bio-security on farms, recreational use of farmland, and crime on farms and within rural communities. She co-edited *Crime in Rural Australia* published by Federation Press in 2007.

Adrian Barton is a Senior Lecturer in Public Policy at the Plymouth Business School. His main research interest lies in interface of drug and health policy in response to drug- and alcohol-related problems.

Walter S. DeKeseredy is Professor of Criminology, Justice and Policy Studies at the University of Ontario Institute of Technology and is the author of 14 books and over 100 journal articles and book chapters. He has been awarded The Institute of Violence, Abuse and Trauma's Linda Saltzman Memorial Intimate Partner Violence Researcher Award in 2008; the 2004 Distinguished Scholar Award (jointly) from the American Society of Criminology's (ASC) Division on Women and the Lifetime Achievement Award from the ASC's Division on Critical Criminology in 2008.

Joseph F. Donnermeyer is a Professor in the Rural Sociology Program, School of Environment and Natural Resources, The Ohio State University. He is an international expert on crime in the context of rural peoples and communities.

Molly Dragiewicz is Assistant Professor of Criminology, Justice and Policy Studies at the University of Ontario Institute of Technology in Canada. Her book *Equality With a Vengeance: Men's Rights Groups, Battered Women, and Anti-feminist Backlash* is forthcoming from Northeastern University Press' Crime and Law series.

Nicholas R. Fyfe is Professor of Human Geography at the University of Dundee, Director of the Scottish Institute for Policing Research and Fellow of the Scottish Police College. His current research looks at the role of activism in the delivery of community safety, police responses to missing persons and the policing of wildlife crime.

Aisha Gill is Senior Lecturer in Criminology at Roehampton University with research interests in criminal justice responses to violence against black and ethnic

Rural Policing and Policing the Rural

minority women. She has served on government working parties on 'honour' killings and forced marriages. She is currently co-editing a book entitled *Forced Marriage: Introducing a Social Justice and Human Rights Perspective.*

Daniel Gilling lectures in criminology and is Director of the Law and Criminal Justice Centre at the University of Plymouth. His research focuses particularly upon the areas of community safety and policing, and upon issues of governance, particularly in terms of partnership working, engagement with communities, and 'boundary work' between criminal justice and social policy.

Keith Halfacree is a Senior Lecturer in Human Geography at Swansea University. His research interests are many and varied but focus on a number of overlapping areas: discourses of rurality in the global North, radical rural futures, human migration, and what he loosely terms marginal geographies.

Matthew Henry is a Senior Lecturer in the Planning Programme at Massey University. Currently his research is dominated by his involvement in a Marsden funded project, 'Biological Economies: Knowing and Making New Rural Value Relations', which is exploring the ongoing transformation of economic value and values in New Zealand's Central Otago and Hawke Bay regions.

Zoë James is Senior Lecturer in Criminology at the University of Plymouth. Her research interests are in policing, Gypsy and Traveller issues, organised deviance and qualitative research approaches, specifically considering the plural nature of contemporary policing provision applied to nomadic communities via community and public order policing initiatives.

Craig Johnstone is a Senior Lecturer in Criminology in the School of Applied Social Science at the University of Brighton. His research interests centre around the control of behaviour in public spaces through both video surveillance and recent public policy innovations concerned with regeneration and anti-social behaviour.

Andrew Lemieux is a post-doctoral Research Fellow at the Netherlands Institute for the Study of Crime and Law Enforcement in Amsterdam. His research interests include wildlife crime, visitor crime and the risk of violence in everyday activities. Much of his understanding about poaching in Africa was obtained during fieldwork with the Uganda Wildlife Authority.

Rob I. Mawby is Professor of Criminology and Criminal Justice at the University of Gloucestershire. He is editor of *Crime Prevention and Community Safety: An International Journal,* and has published numerous articles and books, the most recent being *Policing Across the World: Issues for the Twenty-first Century* (London: UCL Press 1999), *Burglary* (Willan Publishing 2001) and *Burglary*

(Ashgate 2007), and co-authored *Police Science Perspectives: Towards a European Approach* (Verlag fuer Polizeiwissenschaft 2009).

Daniel P. Mears is a Professor at Florida State University's College of Criminology and Criminal Justice. He conducts research on a range of topics, including super-max prisons, sentencing, juvenile justice, public opinion, and re-entry. In *American Criminal Justice Policy* (Cambridge University Press 2010), he argues that an evaluation research framework can create greater accountability and increase the effectiveness of the criminal justice system.

Christian Mouhanna is a permanent Researcher at the Centre for Research in the Sociology of Criminal Law (CESDIP, CNRS). He has worked for 15 years on the relationship between the police and public, as well as strategies of policing. His main publications are related to the French police, the French gendarmerie, French justice and policies of security and criminal justice.

Claire Palmer is a Commissioning Manager for the NHS in Bournemouth and Poole. She worked with the University of Worcester as a Research Officer and, subsequently, for Community Safety Partnerships and Drug Action Teams in the English West Midlands and South West where she undertook needs analyses and service reviews.

Alison D. Reeves is a Reader in Environmental Science, Geography Department, University of Dundee. Her current research interests are: NMR analysis of pollutant related residues in sediments, sewage contamination in estuaries, density gradients and pollutant mobility in estuarine environments, disciplinary interactions: the impacts of disciplinary commitments on diffuse pollution policies.

John Scott is Associate Professor in Criminology, and Convenor of Sociology and Criminology at the University of New England, Armidale NSW. His main research interests are deviance, crime and social control, sexuality and gender, and health and illness. He was a co-editor with Barclay *et al.* on *Crime in Rural Australia* published by Federation Press in 2007.

Greta Squire is a Lecturer in Criminology at the University of Brighton. She is currently completing her Doctoral Theses looking at the role of the voluntary and community domestic violence agencies within the Crime and Disorder Reduction Partnership framework. Her research interests include domestic violence; rurality and partnership working, faith and public policy, gender and the night time economy.

David Storey is a Senior Lecturer in Geography at the University of Worcester (UK) with research expertise on rural social issues, rural development policies and territory, place and identity. He has investigated community-based partnership responses to rural developmental problems, as well as the use of local heritage in rural place promotion. He has undertaken research for a range of local authorities and voluntary bodies.

Michael Woods is Professor of Human Geography and Director of the Institute of Geography and Earth Sciences at Aberystwyth University. His research focuses on rural politics, governance and the engagement of rural areas in globalisation. His publications include *Rural Geography* (Sage 2005), *Contesting Rurality: Politics in the British Countryside* (Ashgate 2005), *New Labour's Countryside: Rural Policy in Britain since 1997* (Policy Press 2008) and *Rural* (Routledge 2010).

Richard Yarwood is a Reader in Geography at the University of Plymouth. He has a wide range of research interests in rural geography and has undertaken work on policing and the emergency services in UK, Australia and New Zealand.

Chapter 1

Introduction

Rob I. Mawby and Richard Yarwood

'What you like to be when you grow up?' a teacher asked her pupil.

'I'd like to follow in my father's footsteps and be a policeman,' he answered.

'Was he a policeman too?' she questioned.

'No,' came the reply, 'he was a burglar'.

If studies of policing have been on the edge of geographical investigation (Fyfe 1991, Herbert 2009), then studies of *rural* policing have fallen off the edge of many research agendas (Moody 1999). Despite renewed and sustained interest in rural studies (Cloke *et al.* 2006), rural policing has received little attention from social scientists (Dingwall and Moody 1999; Yarwood 2001). Rather like the boy in the old joke, academics with an interest in rural policing appear to be following in the tracks of other researchers rather than forging paths on their own.

Yet a focus on rural policing can reveal much about rural society. *Policing* is a broad concept that refers to 'an intricate, almost unconscious network of voluntary controls and standards among people themselves and enforced by people themselves' (Bowling and Foster 2002, 981). Policing is how the police, public and other agencies regulate themselves and each other according to the dominant ideals of society. This can be formally, perhaps through the ever-growing spectrum of policing partnerships in neo-liberal countries, or informally through the performance and enforcement of moral codes and values. Within these broad policing frameworks, it is important to distinguish between demands to reduce crime and demands to exclude activities or people that are threatening to elite rural ideals. It is therefore crucial to realise whose standards are being policed and by whom (Bowling and Foster 2002; Herbert 2009). To achieve this, it is necessary to understand how rurality and criminality are socially constructed and the way that policing affects, and is effected by, these ideals.

This book draws on international, inter-disciplinary perspectives to examine these issues in a range of rural localities. Its authors are drawn from the ranks of geography and criminology. Geographers have a tradition of studying rural society (Cloke *et al.* 2006) yet policing has been 'conspicuously absent from the landscapes of human geography' (Fyfe 1991, 249; but see Herbert 2006, 2009; Yarwood 2007a). Hence the chapters written by geographers in the book reflect a strong engagement with the concept of rural space and who controls it. Their concerns focus more on who is being policed and the implications of this for rural society. By contrast, criminologists focus more directly on those who are engaged in policing, especially the police themselves. Many of their chapters provide rich empirical assessments of policing methods and techniques in rural spaces. Taken together, perspectives from geography and criminology combine to provide important theoretical and empirical views on policing.

The book is organised into two sections. The first examines who is policing rural areas, both in the UK and further afield, and the second examines the nature of rural policing by considering, on the one hand, the policing of rural space and, on the other, how ideas of rurality are regulated.

Part I: Rural Policing

The first section focuses on who polices rural areas by considering the nature of the public police in rural areas in a range of western industrial societies. As policing is conducted by a variety of agencies in addition to the public police, the second group of chapters in this section considers the nature of the policing mix.

In Chapter 2, Rob Mawby offers a broad-brush comparative overview of rural police systems. The nature of the public police varies between societies, but equally the nature of the public police often varies within societies, with differences between rural and urban areas sometimes marked. However, although the Anglo-American image of the rural police is one that has frequently drawn on a perception of the rural as idyllic, and consequently favoured rural police as a model to which urban areas should aspire, he argues that elsewhere rural areas are often policed very differently and indeed that there are significant contrasts between Britain and the USA. The latter exemplifies a system where rural areas commonly have their own autonomous police organisations. In some other countries, such as France, exemplar of a continental police system, and Canada, where the nature of policing was influenced by its former colonial status, rural areas are under the responsibility of a central police agency that tends to be more militaristic than its urban counterparts. Elsewhere, in Australia and England and Wales, regional police systems incorporate the urban and the rural, but in contrasting ways. Mawby argues that the nature of rural police systems depends to a large extent on the social and political circumstances that underpinned the formation of the police, and the extent to which rural issues created particular problems for ruling elites. However, differences between the current situation in Australia and England demonstrate the additional importance of geography, where the sheer size of rural districts and sub-districts in Australia's regional structure restrict the influence of the urban on the rural that is increasingly apparent in Britain.

Chapters 3 to 5 provide more details of the nature of rural police systems within this group of countries. In Chapter 3, Joseph F. Donnermeyer and his colleagues also emphasise the sheer size of Canada and the USA. In each case a large majority of the population is urban-based, with the remainder spread across most of the land mass. Donnermeyer *et al.* note a number of similarities between these rural areas, containing as they do affluence and poverty, growth and decline, and populations of long-standing residents and commuters. They also note the extent of crime and disorder problems, often hidden from view at a time when a decreasing share of resources are being allocated to rural policing. But in other respects, the nature of policing by these close neighbours differs markedly, with the USA providing much more localised services.

The policing of Indigenous peoples is another theme discussed by Donnermeyer *et al.*, and it is one reiterated by Elaine Barclay and her colleagues in the context of Australia. Again, the sheer size of the country, and the implications of policing scattered rural populations, are underlined. But Barclay *et al.* also highlight the complexity of social relations in rural Australian and how this impinges on police work. One particular feature of rural Australia is the presence of Indigenous peoples. Echoing Hogg and Carrington (2006), Barclay *et al.* depict these original inhabitants of the rural as 'outsiders' within their own territory, a problem to be policed rather than a public to be served. Thus, they argue, rural spaces should not be seen as homogenous entities, but instead as diverse and pluralistic settings with competing normative communities, within which the Aborigine population has traditionally received low priority.

The continuing influence of history is also evident in Chapter 5, where Christian Mouhanna considers the evolving position of the *gendarmerie* in rural France. Following Mawby, Mouhanna draws a distinction between the Police Nationale, responsible for policing the cities, and the *gendarmerie*, traditionally responsible for policing the countryside, and acknowledges the more militaristic features of the latter. However, in an insightful insider critique, he argues that in reality the *gendarmerie* has traditionally provided a service more akin to community policing than their urban counterparts. There were, he suggests, at least three reasons for this. Firstly, the organisation of the *gendarmerie*, spread as it is across the country, meant that the granting of local autonomy to territorial squads based in rural districts was, in practice, inevitable, making the *gendarmes* into paramilitary 'street level bureaucrats' (Lipsky 1980). Secondly *gendarmes* had to live in the district where they work, making them – and their families – a part of community life. Thirdly, the lack of resources meant that they were pressured into building good relationships with better resourced local agencies. However, Mouhanna argues, changing expectations among *gendarmes* and their families, and the imminent merging of the *Police Nationale* and *Gendarmerie Nationale*, mean that this, possibly unique, relationship is under threat.

While these four chapters focus upon the public police, there is, as Rob Mawby notes in Chapter 6, a clear tendency in western industrial societies towards a greater reliance upon policing alternatives, a trend towards plural policing (Crawford *et al.* 2005) or multilateralisation (Bayley and Shearing 2001). This is particularly so in countries such as the USA, Canada and Britain, less so in some European countries like France (Jaschke *et al.* 2007; Jones and Newburn 2006). Many of these policing alternatives are provided by the private, for-profit sector, although the use of volunteers and efforts at community self-policing are also common. However, while a number of authors focus on the expansion of plural policing, none address the broader spatial components of this shift. Mawby argues that in the

British Isles policing alternatives that engage the public as volunteers, such as the Special Constabulary, have developed successfully in many rural areas but have been more problematic in cities, and that other forms of multilateralisation, such as the introduction of police community support officers (PCSOs), complement these in an urban setting. Consequently, while plural policing is a feature of both urban and rural areas, the components of this policing mix vary markedly between city and countryside.

In Chapter 7, Daniel Gilling steps back a level and considers the governance of community safety in rural areas. He notes that while a raft of policy measures over the last 25 years have focused upon shifting the emphasis in the policing of crime and disorder – from crime control by the police to crime and disorder reduction through partnership work – government emphasis has been directed at urban, particularly metropolitan, problems. As a result, rural crime has been trivialised or marginalised. Gilling suggests that key players at local government and service provider level have to a certain extent collaborated in this process by either denying that there is a rural problem, blaming the problem on outsiders, or seeing the problem as one of fear rather than risk: representations he describes as the 'Idyllic Countryside', the 'Endangered Countryside' and the 'Frightened Countryside', respectively. He then offers as a more radical alternative of the 'Deprived Countryside', which he sees as a more constructive representation that might underpin rural policing strategies.

The idea that rural crime control has received lower priority than its urban counterpart gains credence from Craig Johnstone's account in Chapter 8 of the development of CCTV initiatives in rural areas. Indeed, in discussing the ways in which rural stakeholders justified a need for CCTV, Johnstone reiterates the three common representations identified by Gilling: the 'Idyllic Countryside' (and the need to keep it that way); the 'Endangered Countryside' (and the need to deter criminals from commuting there to offend); and the 'Frightened Countryside' (and the need to reassure local people). Additionally, he notes the political considerations that underpinned this: once some small towns installed CCTV there was pressure on stakeholders in others to follow suit, both to avoid crime displacement to their patch and to demonstrate to their constituents that they were taking local crime problems seriously. However, there are generic difficulties in translating what was initially an urban initiative to the countryside. For example, the public space covered by what is commonly a handful of cameras is limited and cameras are difficult to monitor. This may mean that a central monitoring point covers cameras located in a number of towns, almost as if CCTV has replaced the village bobby located at a distance from divisional HQ. But in this case the high-tech substitute lacks both the personal touch and the ability for rapid response. This leads Johnstone to speculate, echoing Gill and Spriggs (2005), as to whether CCTV is an appropriate response to crime and disorder in rural Britain, or a top-down initiative imposed by a metropolitan-focused central government.

While both Gilling and Johnstone discuss partnership working in policing rural crime, this receives more explicit attention in Chapter 9, where Richard Yarwood uses case studies from England and Wales, Australia and New Zealand to assess who is responsible for local policing in both de facto and de jure terms. While, as with CCTV, he argues that rural policing initiatives have been influenced from the top down in Britain and Australia, with the resultant difficulties of translating policy into practice, he points to the New Zealand model as one of bottom-up initiatives.

Part II: Policing the Rural

The chapters in Part II then focus on the nature of crime and disorder in the countryside and the ways in which this is policed. While these chapters offer an eclectic mix, they epitomise two distinctions that might be drawn: between 'traditional' rural crimes, such as agricultural crime and poaching, and crimes that are commonly associated with urban life, such as drug misuse and domestic violence; and between problems associated with local residents themselves and those deemed to be 'outsiders', reiterating Gilling's representation of the 'Endangered Countryside'. In each case, policing may vary accordingly. Thus 'rural' crimes may evoke specialist units, whereas conventional crimes may pose particular policing problems where rates are lower, or incidents are less visible, than in the city. And the police may find it less problematic to deal with outsiders than insiders where their relationships with local stakeholders are less likely to be undermined.

The latter point is nicely illustrated in Chapter 10 where Mike Woods considers the policing of rural protest. He notes that early protests reflected a 'displacement of urban politics', where protest was located in the countryside because that was where nuclear power stations or military bases tended to be. Later protests against road construction and by animal rights activists or hunt saboteurs, were similarly portrayed as urban incursions, where policing agents could rely on the rural vote to support their actions, reflected in *The Criminal Justice and Public Order Act 1994,* that essentially protected rural people from outside incursions. However, from 1997 onwards, farmers' protests and pro-hunting groups, often coalescing in the Countryside Alliance, posed a different set of problems for the police, who needed to deal with 'protest from within', and initially at least adopted the mantra of policing by consent. However, as more radical offshoots of the Countryside Alliance opted for direct confrontation, police tactics have changed, unsettling the dynamics of police-community relations in rural areas.

In contrast, Chapters 11 and 12, by Keith Halfacree and Zoë James, address travellers, arguably the 'folk devils' (Cohen 1972) of rural societies. While both chapters focus on the UK, as Halfacree demonstrates, there has been longstanding prejudice against travellers in a range of countries. Despite a romanticised portrayal of the travelling life[1] and a need for some of the specialist or seasonal work performed by travellers, they have been seen to pose a threat to the countryside in three ways: they disrupted the 'predominant spatial practices', especially through 'disrespect' for private property; they challenged the everyday lives of local people, presenting an alternative way of living; and they challenged the rural idyll. In England and Wales, *The Public Order Act 1986*, *The Criminal Justice and Public Order Act 1994* and *The Anti-Social Behaviour Act 2003* explicitly target travellers, the former removing the requirement contained in *The Caravan Sites and Control of Development Act 1968* for local authorities to provide sites for travellers. Halfacree's analysis is complemented by James, who identifies key issues of control and policing as drawn out of a national sample of local authority reports on the needs of Gypsies and Travellers. James argues that the relationship between travellers and local law-enforcement agencies is predicated by the fact that most occupy unauthorised, and consequently illegal, sites, and so hostile reaction from the 'settled community' and the consequent call to evict sets the policing agenda. Not surprisingly, then, travellers lack trust and confidence in the police, who are associated with eviction and enforcement activity. When overt conflict occurs, the police utilise a range of public order policing tactics to manage travellers, resulting in a further reduction in their confidence in the police and authorities generally, who are seen as adopting the settled community's perception of travellers as requiring controlling and managing, rather than viewing them as citizens with a right to a comprehensive range of police services.

Increasingly, urbanised societies such as Britain make much of the idyllic nature of 'the country' and the benefits of 'rural living'. Chief amongst those alleged benefits are the apparent problem-free nature of rural settings and the absence of many of the inherent problems associated with city living. Whilst it is true that there are differences in day-to-day living between rural and urban settings, it is not true that rural living is without problems. There is, however, a tendency for rural problems to be under-explored and under-reported and for 'rurality' to be conceived as a holistic geography, when this is clearly not the case. Chapters 13–14 illustrate this by describing crimes that are conventionally associated with the urban, rather than the rural. Drawing on empirical data, Adrian Barton, David Storey and Claire Palmer seek to address these misconceptions of the idealised, problem-free, holistic nature of the rural idyll by exploring the nature and extent of illicit drug use in two rural settings (Cornwall and Herefordshire), and the manner in which services for drug users are provided. Noting that on a national level drug misuse among younger people appears to be at least as common in rural as in urban areas, they use their research to map out local drug scenes. They argue, following the discussions of plural policing, that the traditional police play a limited role in controlling drug misuse, and that users are more likely to experience close relations, and be policed, by a range of health and welfare agencies operating

1   A US equivalent can be found in the work of singer/songwriters such as Woody Guthrie, Bob Dylan and Tom Paxton.

within local DAATs. To a certain extent, drug misuse in rural areas, despite its public location, is relatively invisible. The same applies to an even greater extent in the case of domestic violence (Hogg and Carrington 2003). Drawing upon empirical data gathered within Oxfordshire and the South West, Greta Squire and Aisha Gill argue that living within a rural setting will often exacerbate the effects domestic violence has upon the victim by geographically isolating them due to poor transport, and institutionally isolating them from support through the lack of suitable agencies tasked with dealing with the needs of the survivors. This is often compounded by the fact that 'policing' domestic violence extends beyond the police service: effective service provision is often about including a number of welfarist based organisations, which can lead to problems of service co-ordination, ownership and resource management.

In contrast, the final group of chapters deal with what might traditionally be considered as typically rural policing concerns. Chapters 15–16 consider wildlife crimes that are perhaps unique to rural places. The policing of wildlife and environmental crime are crucially important in terms of the protection of natural heritage and the social and economic activities that depend on the natural environment, yet our understanding of how such policing is carried out remains rudimentary. Drawing on interviews with police wildlife crime officers (WCOs) and wildlife crime coordinators (WCCs) in Scotland's eight police forces, Fyfe and Reeves focus on the resources, roles and responsibilities of WCOs and WCCs, the interplay between reactive and proactive policing engaged in by these police staff, and their perceptions of the impact of their activities on wildlife and environmental crime in Scotland. In the next chapter, Andrew Lemiuex extents this analysis across state borders and considers the geographies and policing of elephant poaching in and out of sub-Saharan African. This chapter illustrates the dangers of focusing exclusively on rural crime or rural policing. Paralleling the situation *vis-à-vis* drug crime, poaching involves three processes: acquiring the illegal product, transporting it to viable markets, and selling it. These processes clearly transcend the rural locale, and indeed international boundaries, and highlight the importance of cooperation between policing agencies that may have very different priorities.

Finally consideration is given to agriculture, an activity that is associated closely and uniquely with rurality. Although only a small corpus of empirical work on agricultural crime has been conducted over the past quarter century, their results point a high cost to farm victimisations and distinctive correlations between physical and topographical features of agricultural operations and the location of crimes. The chapter by Elaine Barclay and her colleagues considers the extent and pattern of agricultural crime. It utilises an ecological perspective to identify vulnerability to various crimes and reviews attempts by the police prevent them. By contrast, Matthew Henry considers how farmers are themselves policed by a range of agencies and technologies. The chapter situates the rise of these technologies within what Foucault termed a 'biopolitical' modality of power through which the state has become increasingly concerned with issues of 'improvement' in relation to the dynamics of both population and economy. Couched in these terms the chapter explores the webs of regulation that have shaped farm production in New Zealand. It provides an apt end to the book, noting that plural policing is no recent phenomenon and that it incorporates alternative state agencies as well as alternatives to policing by the state. Additionally, it may be focused on activities very much in keeping with what is understood and expected as rural. Contrary to the fears expressed at the start of this introduction, rural policing is far wider than providing protection for farmers by the state.

PART I
Rural Policing

Chapter 2

Rural Police: A Comparative Overview

Rob I. Mawby

Introduction

Rural policing is often idealised as a community-based affair. In the USA, for example, there is an implicit assumption that the police should live, work and be part of a local community (Thurman and McGarrell 1997; Weisheit, Wells and Falcone 1994). In Japan, the urban police station, or *koban*, was modelled on the rural *chuzaisho*, police houses where local officers lived with their families, providing a range of services to 'their' communities (Ames 1981; Bayley 1976, 1994).

However, this view of a rural policing idyll is a culturally bounded image. Police systems were established to counter particular problems that those in power identified as a threat and, as the nature of threat varies between countries, so too do the police. 'The police' can therefore be classified under one of five ideal types, Anglo-American; Continental European; Communist; Colonial; and Far Eastern (Mawby 1990, 1999), according to structure, core functions and accountability. As 'Threat' also varies, quantitatively or qualitatively, between city and countryside, urban and rural policing can vary markedly within these systems.

Thus, in Britain and the USA this may reflect a rural policing idyll yet in other societies with traditions of centralised police systems, rural police were often more militaristic and less publically accountable than their big-city counterparts. In France, Italy and Spain, for example, where two centralised police systems have existed side-by-side, it has traditionally been the more militaristic alternative that has held responsibility for rural policing, with officers routinely housed in barracks and segregated from local influence. In Canada, the militaristic Royal Canadian Mounted Police (RCMP) normally covers the provinces, with local civilian forces more common in the cities.

The following three sections critically compare three alternative models of rural policing, viz: localised policing (USA); centralised policing (France and Canada) and regional policing (Australia, England and Wales).

Local Police Services: The USA

The United States has a multitude of autonomous local police organisations and is in many ways unique (see also Donnermeyer *et al.*, Chapter 3 this volume).

Early European (predominantly British) settlers established rudimentary law-enforcement systems in the emerging towns and cities of the East coast, with inland, more rural systems following later. 'Modern' police systems (Bayley 1985) in the USA were then created in the aftermath of the struggle for independence, initially in cities like New York and Boston (Miller 1977; Monkkonen 1981). From the outset, there was concern that external control from Westminster (London) should not be replaced by internal central control from Washington, with the result that autonomous local law-enforcement systems were encouraged (Sweatman and Cross 1989). Of course, there are also a number of national, or federal, police institutions, perhaps the best known being the Federal Bureau of Investigation (FBI) and Drug Enforcement Agency (DEA), but these are tolerated rather than welcomed (Loveday 1999).

Local police organisations can be categorised under five headings: state, county, city, small town and 'special district' (Bureau of Justice Statistics 2002; Reaves 2007; Sweatman and Cross 1989; Walker 1983; Weisheit *et al.* 2006):

- *State.* Forty-nine states have their own police force, in total employing over 58,000 full-time sworn officers. State forces are commonly accountable to state governors.
- *County.* Most of the counties in the USA have their own police departments, headed by a sheriff elected to run the department for two–four years, who is in turn accountable to the county administrators. In 1999 there were some 3,088 sheriffs departments employing about 185,859 sworn officers.
- *City.* There are about 3,220 police forces in cities of over 250,000 population, employing on average 2,465 sworn personnel.
- *Small town.* The fourth level of local law enforcement operates at the rural level in the small towns. There are at least 13,000 such forces, employing about 438,000 full-time sworn officers, with nearly half (47%) of these agencies located in areas with under 2,500 people.
- *Special districts.* These provide the police services to particular institutions including parks, recreation areas and wildlife sanctuaries. In 2004, there were 205 of the latter employing over 14,000 sworn officers.

In terms of personnel the typical US police officer works for a large urban department but the most common police *organisations* are small-town, independent forces, with about 9,000 departments employing less than ten staff each, including civilians (Reaves 2007: 1–2).[1]

Rural police agencies as epitomise many of the features of the community policing model (Falcone, Wells and Weisheit 2002; Thurman and McGarrell 1997; Weisheit, Wells and Falcone 1994), frequently incorporating a problem-oriented approach (Chaiken 2004). Officers are more likely to be drawn from the community in which they work (Weisheit *et al.* 2006); they are more directly accountable

---

1   For a personal example, see Pierce (2001).

to that community, either directly where sheriffs are elected or indirectly where officers are subject to community pressure (Weisheit *et al.* 2006); they are better integrated into that community (Decker 1979; Sims 1988); they carry out a broader range of service-oriented tasks (Johnson and Rhodes 2009; Lieberbach and Frank 2003; Payne *et al.* 2004; Weisheit *et al.* 2006) and operate as generalists rather than specialists (Galliher *et al.*, 1975; Maguire *et al.* 1991); and they receive more positive approval from local people (Weisheit *et al.* 2006).

There are, however, problems with this. On the one hand, rural officers who are cut off from their colleagues are less able to call back-up (Hoffman 1992). While they are less likely to be assaulted than urban colleagues, when assaults occur they are more likely to be fatal (Weisheit *et al.* 2006). Moreover, whilst isolation from the chain of command might appear to increase discretion and autonomy, the same authors suggest that control by the community might more than compensate for this. More generally, they may find it more difficult to keep separate leisure and family life from work (Pierce 2001; Weisheit *et al.* 2006). On a financial level, rural forces appear to lack some of the technological aids available to their metropolitan counterparts (Donnermeyer *et al.*, this volume; Eastern Kentucky University 2003), although there is some evidence that smaller departments have been successful at subcontracting out specialist work to larger neighbouring forces (Ostrom, Parks and Whitaker 1978). Additionally, economies of scale may make urban forces more economical (Krimmel 1997).

Rural Police in a Centralised Structure: France and Canada

France is, perhaps, the classic example of a Continental (European) police system (Mawby 1990). Traditionally, the police systems on continental Europe have been more centralised and more wide-ranging in their responsibilities than in England or the USA. They may be characterised as structurally, more centralised and militaristic; functionally, more focused on political and administrative tasks; and, in terms of legitimacy, more closely tied to government and less accountable either to public or law (Mawby 1990).

That said, most Western European countries have more than one police force, at least one of which is a centrally based, militaristic force that tends to be responsible for rural policing. In France and the Netherlands, this traditionally ran alongside a number of locally run urban forces. In the Netherlands, concern over lack of central influence over funding (Wiebrands 1990) led to the complete overhaul of the system, resulting in regional forces (Jones 1995) not dissimilar to the situation in England and Wales. In France, amalgamation of urban forces in the post-war period, culminating in the incorporation of the Paris police into the Police Nationale in 1966, resulted in two centralised police systems, the *Police Nationale* to cover all urban area and the *Gendarmerie Nationale* with responsibility for rural France.

The latter has its roots in the *marechaussee*, a highly militaristic organisation that prioritised, at different times, political disorder, vagrancy, and the threat posed by highwaymen (Emsley 1983; Stead 1983) and was ultimately answerable to the Minister of War. The *marechaussee* was reformed in 1791, following the revolution, into the *Gendarmerie Nationale*.

Today the *Gendarmerie Nationale* remains the more militaristic of the major police organisations, located, until 2009 (see Mouhanna this volume) under the Ministry of Defence and with a range of functions including checking firearms and policing the armed forces (Jammes 1982; Stead 1983). While both the *Police Nationale* and the *Gendarmerie Nationale* are accountable to central government, rather than having any direct local accountability, the *Gendarmerie Nationale* has a strong local presence in rural areas, with *brigades* based within most of the *cantons* covering rural France. These are housed in barracks, in many cases with officers' families also living within the barracks, reflecting the militaristic nature of the organisation.

This militarist approach reflects a historic requirement to police against threats of border incursion, attempted invasion and internal insurrections where remote rural areas posed a potential threat. Control of the main routes to the capital was therefore prioritised and a paramilitary police force considered the most effective way of maintaining this. As the nature of the rural threat has changed, it is arguable that the *Gendarmerie Nationale* has also changed to reflect this (see Mouhanna in this volume).

While policing systems in some areas of Canada have been influenced by the French who settled there, police structure in rural Canada was more immediately defined by the British colonial experience and paid town forces were established, as in England, during the 19th century. Yet the need to police vast, under-populated areas of rural Canada required distinctive forms of policing, based on the colonial mode (Morrison 1985), that aimed to control Indian and Inuit populations, protect the territory from foreign incursions and impose order within the goldfields. In the 1850s the colonial authorities in British Colombia were sent a sub-inspector from the Irish Constabulary to establish a similar colonial force there (Chapman 1978).[2]

This was subsequently identified as appropriate to meet the varying challenges to British rule across the vast territories of Canada (see also Donnermeyer *et al.*, this volume). As Morrison (1985: 136) observed, 'nothing establishes sovereignty over an area more clearly than effective policing of it.' Originally known as

---

2   Similarly, in rural areas of the USA that had significant public disorder problems and a perceived threat from labourers, 'aliens' and the like, police organisations were formed that showed striking similarities to the colonial model. The establishment of the Pennsylvanian force in 1905, for example, heralded a new wave of state forces designed to 'crush disorders, whether industrial or otherwise, which arose in the foreign-filled districts of the state' (Reppetto 1978: 130). A parallel model emerged in some southern states where a police priority was to round up escaped slaves (Reichel 1988).

the North-West Mounted Police, this centralised, militaristic body extended its influence in the early 20th century, taking on security and counterespionage services during the First World War and, in 1919, helping to break the Winnipeg general strike (Horrall 1980). In 1920 it was renamed the Royal Canadian Mounted Police (RCMP) and continued to expand, on the one hand taking on additional responsibilities from federal government, on the other hand extending its services in the provinces. By the 1930s it provided police services in the Yukon and Northwest Territories and for six of the 10 provinces; in the post-war period it assumed provincial policing responsibilities in two further provinces and established a foothold in urban policing (Chapman 1978).

Under the 1867 constitution, modified in 1982, policing within each province is a matter for local government to decide, with the result that metropolitan and provincial forces generally coexist. But the RCMP provides three sets of services that impinge on this local blueprint. It acts as a federal police force with national responsibilities throughout the country. It maintains responsibility for policing the Yukon and Northwest Territories and it can be subcontracted to provide police services at provincial and municipal level. In Ontario and Quebec the RCMP's responsibilities are confined to national legislation and policing is provided by metropolitan and, to a lesser extent, provincial forces. However, elsewhere the RCMP has a substantial foothold as the main service provider, especially in rural Canada. In general, then, urban and metropolitan police systems are organised and managed locally, with rural services provided centrally through the RCMP. The latter, with its traditional order-maintenance emphasis and its involvement in a wide range of administrative responsibilities (Dion 1982; Morrison 1985; Weller 1981), appears as the more remote and militaristic alternative. Moreover with its centralised structure it is clear that local accountability has traditionally been restricted, even where its relationship is a contractual one (Weller 1981).

As in France, the historical foundations for this appear to have derived from the different problems faced by government in maintaining order in the countryside, in this case the colonial authority in exerting British sovereignty in the vast rural hinterland. However, it is equally clear that the Canadian government has recognised the advantages of a militaristic centralised body that can be relied on (for example during metropolitan police strikes in the 1920s and 1970s) and which effectively has prevented close scrutiny of political policing, for example with regard to operations against the Quebec separatists in 1972–74 (Dion 1982; Weller 1981). Although it has been forced to change, the fact that central government effectively subsidises the RCMP means that provinces in general, and rural areas in particular, may be willing to sacrifice local accountability for financial savings.

These examples from France and Canada suggest that the US image of rural police forces as more localised, more community oriented and more directly accountable to their publics may be a misleading example from which to generalise. The next section contrasts the Australian and mainland UK experiences, where rural police operate within a regional structure that encompasses both metropolitan and rural areas within the same agency.

Rural Police in a Regional Structure: Australia and England and Wales

Australia

As in Canada and the USA, police forces in Australia were initially established in emerging cities and townships. Consequently, the police structures in those settlements resembled those in England and Wales, with (honorary and then paid) magistrates controlling the police. However, while in Canada a centrally-based paramilitary force was created to manage the vast rural hinterland, in Australia, similar threats to British control were countered through forces managed at state level and emanating from colonial administrators in the major towns (Finnane 1987; Neal 1991; Petrow 2001). Rural concerns centred around at least three problems: the need to control the Indigenous population while opening up their land to settlers; policing convict banditry; and, later, the need to impose order in developing mining communities.

These problems differed qualitatively and quantitatively from those in the urban centres and so the police model adopted there, although not centralised above state level, shared much with the British colonial model. It was paramilitary, given a broad range of administrative and order-maintenance responsibilities, and provided instant 'justice' that was only weakly accountable to the urban-based administrators (Finnane 1987; Richards 2001). In New South Wales, for example, the Mounted Police were responsible for ruthless slaughter of defiant Aborigines (Neal 1991).

While many features of these early policing forms have changed, the power of the individual states meant that Australia's criminal justice system in general and police system in particular are state-based, with the federal police far less significant than in Canada or even the USA. Each of the six states, plus the Northern Territories, has its own police system, with the Capital Territory Police based in Canberra a minor feature. The structure is thus essentially a regional one. However, a small number of regional forces cover a vast, remote country which means that the regional structure is dramatically different from that in England and Wales (see Barclay, Scott and Donnermeyer, this volume).

The overall consequence of this is that rural police officers are often based on their own, with back-up literally hours away. This makes them highly dependent upon support from within the local community and difficult for the police to separate their public and private lives, meaning they – and their spouses – may be on duty continuously (Adcock 2002; Jobes 2002; Moore 1992). Unlike in the USA, where the highly localised system encourages local recruitment, officers are not generally drawn from the areas they police and postings to remote rural areas may be unpopular. This is for social and financial reasons (officers posted to rural areas may find their spouses' jobs compromised) and due to concerns about a lack of suitable housing (Adcocks 2002).

In these circumstances, there is a need for newly posted police officers to establish themselves in their new communities, with support from local people and partnerships (Adcocks 2002). As in other countries, the wide variety of functions undertaken by the police is acknowledged, with the consequential importance for them to operate as generalists (Jobes 2002; Pennings 1999). However, the assumption that they can operate both effectively and appropriately within a set of gemeinschaft relationships is questionable (Scott and Jobes 2007), given that the rural idyll is both gendered and racialised (Barclay, Scott and Donnermeyer, this volume). Hogg and Carrington (2006) have argued that control in rural areas focuses upon control of public space, which in turn centres upon the Indigenous population and a series of public disorder scenarios. This may treat Aborigine women as offenders but ignores them, and white women, as victims of domestic violence. Relatively high rates of recorded rural crime are thus inflated by an overemphasis on public disorder offences and deflated by a reluctance to question rural harmony by taking domestic violence seriously.

The policing of Indigenous people in rural areas has its origins in the suppression of the Aborigine population. Subsequently, police involvement with residential, educational, occupational and parenting impositions, and specifically through the forced removal of Aborigine children from their homes, mean that police relationships with the Indigenous population have often been couched in terms of control of inadequates, troublemakers and undesirables rather than in terms of support for citizens and victims (Barclay, Scott and Donnermeyer, this volume; Cunneen 2001, 2007; Jennett 2001; Mazzerole *et al.* 2003). The high rate of recorded offending by Indigenous people is both a consequence of this and a mechanism of its perpetuation.

England and Wales

The police structure in England and Wales is also regional (Mawby and Wright 2008), but it differs fundamentally from that in Australia. The modern police system is derived from successive Acts through the nineteenth century that resulted in three types of police force: the London Metropolitan Police, county forces and borough police. One key distinction between the rural (county) and urban (borough) police was that the former were managed by local elites, with chief constables generally the younger sons of the local aristocracy, whereas in the towns and cities the local elites ran the Watch Committees that appointed middle-rank managers to head the police under their supervision (Wall 1998).

This highly localised system remained until the 1940s, when nearly 200 separate local forces existed in England and Wales. These were reduced to 43 by the mid-1970s. The system was broadly regional, based on metropolitan areas and counties (or pairs of counties) and while the government has mooted the further amalgamation of smaller neighbourhood forces it has, for the time being at least, shelved a move towards either a national force or one based on a much smaller number of super-regions. Thus while there are some forces that are almost entirely metropolitan, and the Home Office has variously produced 'families' of forces to identify the similarities and differences between forces, most forces contain a mix of urban and rural areas (Aust and Simmons 2002). Within constabularies, areas are further broken down into Basic Command Units (BCUs) that often contain urban and rural environments so that the city and country are essentially policed by the same organisation.

Nevertheless, we might expect some differences in the day-to-day nature of policing in urban and rural parts of the same force. Writing about research conducted in the 1950s, Cain (1973) underlined the extreme differences that existed at that time. In urban areas, the police operated as a team. They lived and worked in close proximity to their colleagues, with the result that a strong occupational subculture existed that, at times, encompassed their families. In contrast, in rural areas officers were isolated from their colleagues and more dependent upon their neighbours; and, living and working in the same 'patch', their families were accorded a distinctive status within the community.

The situation is very different in the 21st century. Improved technology, particularly motorised transport and high-tech communications systems, have transformed rural policing. At the same time, police officers no longer live in constabulary-owned police-housing (a form of 'tied housing'), and are generally deployed from a central station. Research in Cornwall, a largely rural county in England with no towns registering more than 30,000 people, revealed that the public were reluctant to accept that the police were fully integrated in their local communities, seeing them as police officers first and locals second (Mawby 2004, 2009). They identified difficulties accessing the police, possibly reflecting the closure or part-time openings of local substations and the rerouting of phonecalls to a force-wide call centre in Exeter, the neighbouring county on the eastern edge of the force area. Interviews with police officers also confirmed that the contrast between urban and rural police that Cain (1973) described some 30 years ago is no longer so evident (Mawby 2004, 2009). Advanced communications systems and a reliance on patrol cars have meant that police officers are no longer isolated from their colleagues. At the same time, better transport and the closing of rural substations means that they are no longer 'marooned' in the areas where they work, enmeshed in a local structure where their identity as a police officer balances precariously with other local identities. Nevertheless, citizens in our survey still tended to distinguish between rural and urban police, seeing some advantages in the former and hinting at a wider perception that rural police are, or should be, viewed as a part of the local community.

The indications here, then, are that even in a comparatively rural area like Cornwall, the police operate in broadly similar ways to their colleagues based in urban and metropolitan areas. To a greater extent than in Australia, where the area covered by rural police units is vast, new technology has served to draw rural officers closer to their urban counterparts and make them less reliant upon local residents.

Rural Police in Context: Reassessing the Rural Idyll

Even a limited analysis of the way the police are organised and operate in rural areas is sufficient to demonstrate that the US experience is scarcely typical. By allowing local autonomy from the outset, the US constitution gave local governments in rural areas the opportunity to create, maintain and control their own police institutions. In some cases local control was superseded by national agencies or other forces covering larger regions; in other cases, small local agencies have entered into cooperative agreements with larger agencies where there is a need for manpower cover or specialist expertise. Generally, though, the picture is one of local agencies addressing local problems that might be specific to that area.

However, while the granting of local autonomy can be understood in terms of the struggle for independence, in other countries very different historical imperatives underpinned the type (or types) of police institution created and, specifically, the type of policing deemed appropriate in rural areas. On continental Europe, Communist societies, and the British Empire, threats to the established order, either internal or external, meant that more centralised control of policing was considered necessary. In countries like France, this was particularly the case in the countryside, with the consequence that a militaristic form of policing was established to cover rural France. Not surprisingly, this has been modified in recent years, a point argued by Mouhanna in this volume, but it would make a change of seismic proportions to make it similar to its US counterpart.

Although not all colonial police systems were centrally controlled, most were characterised by militaristic qualities. Among the 'old' British colonies, Canada provides an example of a mixed system, with local, civilian police commonly operating in the metropolitan areas and the national police system, the more militaristic RCMP, covering most of rural Canada. Thus while urban police in the USA and its neighbour share many similarities, the policing of rural areas differs markedly. Personnel are not, for example, recruited locally in rural Canada, and the concept of community policing has tended to be stronger in urban Canada.

As another 'old' colony, Australia shared many of Canada's law and order problems, albeit the incursion of foreign powers was less important. However the need to subject the indigenous population and maintain order in rapidly industrialising rural areas such as the goldfields were made more difficult in a country where a significant proportion of the population was former, or escaped, convicts. Regional autonomy operated against the creation of a significant federal force, but rural policing based on the colonial model was run from the colonial administrative bases in the major cities. Again, modern police systems differ significantly from the original model, but it is easy to see echoes of the earlier form: for example in the intra-state centralisation of the police and in the control of the Aborigine 'problem'.

The police in England and Wales, as in Australia, operate within a regional framework, but historically shared more features with the USA than with its other colonies, for which the British created an arguably different control apparatus. In some respects, the differences between rural policing in England and Wales and Australia reflect these different traditions. Thus even after the amalgamation of a myriad of local police agencies into 43 regional forces, the picture painted by Cain (1973) mirrors that of much of the US literature. However the changes that have taken place since then owe much to new technologies that have in effect drawn the rural into the urban. In Australia, with immensely greater regions and a more sparsely populated countryside, the impact of technology has been more muted. To take but one example, when a WA police officer is transferred from Perth to one of the eight rural districts, the move is life-changing, involving moving partner, possibly to a new job, and children to a new school. In complete contrast, when a police officer serving with the Devon and Cornwall Constabulary is transferred from urban Plymouth to Cornwall it is possible to remain living in the city and commute to work. And the latter are equally in much closer contact with their colleagues and conversely less dependent upon local people. It is consequently scarcely surprising that respondents from our surveys saw differences, but not excessive differences, between 'their' police and those working in metropolitan areas.

This may be a disappointment for those viewing with nostalgia a rural policing idyll. But is the US model of local rural police agencies one that should be developed elsewhere? Many US commentators are in little doubt (see for example Simms 1988), with their local roots, community and problem-oriented philosophy, and service orientation commended, and community accountability praised. In Britain too, Loveday and Reid (2003), writing in response to the Conservative opposition's brief flirtation with the idea of smaller, directly accountable police services, have argued in favour of this one aspect of the Americanisation of the British police. However, both US research (see also Donnermeyer *et al.*, Chapter 3 this volume) and findings from other countries suggest the need for caution.

One issue relates to the cost-ineffectiveness of small rural forces, an argument contested by Loveday and Reid (2003) but underlined by some US research (Eastern Kentucky University 2003; Krimmel 1997). Another argument relates to the dangers of generalism and the advantages of being able to offer specialist expertise, in relation to crimes as different as domestic violence and burglary.

Perhaps most importantly, though, the debate around the advantages of having policing carried out by local people, accountable to local constituencies, is a contentious one. On the one hand, advocacy of a system of elected sheriffs as in much of the USA is dented where in the same country the election of the local judiciary has routinely been highjacked by a powerful law and order lobby advocating less rights for defendants and more severe sentencing, and seems to imply that justice equates to majority opinion. Similarly, the fact that police are local people is not necessarily a strength, where local people may retain local prejudices. The fact that the Japanese police system, allegedly one of the most community-oriented in the world (Ames 1981; Bayley 1976, 1994), has been wary of leaving officers in one post for too long, warns of the dangers of police becoming over-dependent on local people. This is particularly the case because, contrary to rural mythology, rural policing rarely takes place in homogeneous gemeinschaft communities. As the Australian example illustrates, police officers will identify with particular constituencies within 'their' community, in this case white males. In Northern Ireland, an extreme divided community, policing by a largely Protestant police force of a population with a slight majority of Protestants created neither a fair nor effective force. In the USA itself, ensuring that the ethnic composition of the police in the big cities broadly paralleled that of the populous led to endemic corruption. The example of *Diskworld* notwithstanding, it is more important for police officers to be representatives of justice than to be representative of the population. Equally, while it is possible to identify a model of policing that is most appropriate for an area, it would be naïve to advocate some mythological rural police system as a panacea to today's police problems.

Chapter 3

Policing Rural Canada and
the United States

Joseph F. Donnermeyer, Walter S. DeKeseredy and Molly Dragiewicz

The challenges of rural policing in both Canada and the USA appear easy to recite: large geographic areas that are difficult to patrol, limited resources for the hiring of sufficient, trained personnel and a general neglect by provincial/state and national leaders who (often) mistakenly believe that crime in rural communities is comparatively much lower than in urban centres (Weisheit *et al.* 2006). Yet, issues of policing in rural Canada and the USA are defined less by geographical distance and location, and more by structural and economic factors that shape rural spaces.

There seems to be two faces to rural Canada and the USA: some rural areas border on conditions found in the poorest countries of the world, while others approach idealised notions of what most citizens of both countries would define as the 'good community' (Lyson and Tolbert 2003). Large swaths of rural areas are awash in poverty, and social, economic and political inequality, racism, and community norms that are surprisingly tolerant of crime, especially various forms of interpersonal violence (DeKeseredy *et al.* 2009; Harris and Worthen 2003; Jensen *et al.* 2003; Sherman 2005). Marijuana and methamphetamines undergird local economies in some of these places (Donnermeyer and Tunnell 2007; Weisheit 2008). In contrast, there are some very affluent rural communities in Canada and the USA that enjoy an expanding population base and various forms of residential and commercial development.

Yet, this dichotomy oversimplifies rural society and limits debate about rural crime (Donnermeyer 2007a, 2007b). Most rural communities are a complex mix. Persisting poverty, racial segmentation, intolerance of alternative lifestyles and uneven enforcement of laws across various peoples and social classes exist side by side with strong ties of personal relations based on family, neighbourhood, volunteer groups and other forms of tight-knit informal networking (Freudenburg 1986). In a sense, there is an underclass, often hidden, in most rural communities of Canada and the US (Rephann 1999; Smith 2008). These complex rural realities shape both the nature of crime and its policing in rural places.

The purpose of this chapter is to describe the nature of policing in rural Canada and the USA. We examine the social organisation of policing and how broader social structural issues reflect the rural realities of policing in Canada and the USA. We conclude with recommendations for more comparative and critical scholarship on police studies in the rural context.

Crime and Policing in Rural Canada

Cultural Background

Since Canada's colonial beginnings, its economy, culture, and criminal justice system have been heavily shaped by foreign influences (DeKeseredy and MacLean 1993; Grabb 2004). The Canadian justice system 'shares a common legal heritage with England' (Terrill 2007: 203) but is increasingly influenced by laws, policies and practices that are based on US models (DeKeseredy 2009; Jones and Newburn 2002; Newburn 2002). This is set against a widely held view that rural Canada is relatively safe, at least compared to the USA.

Jeffrey Simpson (2000: 95) points out: 'Canadians prefer to think of their country as virtue incarnate, its cup of tolerance running over. They endlessly recycle the cliché about Canada the "peaceable kingdom" in large part because it makes them feel good about themselves. Canadians are peacekeepers abroad, peaceful citizens at home.' Yet rates of domestic violence and sexual assault (Cross 2007; DeKeseredy 2009; DeKeseredy *et al.* 2007; Sev'er 2002), as well as the exclusion and repression of Aboriginal people in Canada (Restoule 2009), suggest that Canada is not necessarily a 'kinder, gentler nation'. Although frequent and positive comparisons with the USA constitute 'a fact of life' (Grabb and Curtis 2005), 'it is impossible to say', other than homicide whether the overall Canadian crime rate is lower than the USA's.

Further, many people assume that crimes seldom occur in rural Canadian communities (DeKeseredy *et al.* 2009). This belief is heavily fuelled by the media, lay conversations and even criminological research, which typically focuses on urban lawbreaking (Donnermeyer *et al.* 2006). According to *Toronto Star* newspaper reporter Theresa Boyle (2007: A3), 'conventional wisdom holds that the big, bad city is the root of all evil. Small towns are supposed to be peaceful and serene.' Yet a recent Statistics Canada (2007) study found that crime is certainly not restricted to large cities in Canada. Analyses of 2005 police-reported data reveal that the homicide rate of 2.5 per 100,000 people in rural areas was higher than the rate of 2.0 in large urban areas and the rate of 1.7 in small urban ones. This pattern also remained constant over the past 10 years. Furthermore, Statistics Canada's 2004 General Social Survey found no significant difference in rates of physical assault experienced by urban and rural residents (Gannon and Mihorean 2005).

The Structure of Rural Policing

Policing in Canada is divided into four segments: federal, provincial, municipal/regional, and First Nations. The duty of rural policing falls largely to the Royal Canadian Mounted Police (RCMP), Canada's federal police force. The RCMP grew out of the North West Mounted Police (NWMP) force, which was established in 1873 to patrol the Northwest Territories (Royal Canadian Mounted Police 2008c). Early on, primary concerns of the RCMP in these rural and developing areas were: addressing liquor sales, managing conflict between indigenous people and Canadian settlers, collecting customs dues, and performing policing duties. Later, NWMP forces were deployed to maintain order during the gold rush and to prevent American encroachment on newly acquired Canadian territory (Royal Canadian Mounted Police 2008c).

In 1920, the NWMP was merged with the Dominion Police force which served eastern Canada at that time (Royal Canadian Mounted Police 2008c). In addition to enforcing federal laws and protecting border security and other federal functions, the RCMP provides police service under contract for all Canadian provinces except Ontario and Quebec. It also provides police service to the Northwest Territories, Nunavut, Yukon, and 197 additional municipalities under agreements between the RCMP and local communities (Royal Canadian Mounted Police 2008a).

Each province and territory has its own policies for providing policing to rural areas. For example, according to British Columbia's Police Act (1996), municipalities larger than 5,000 people are required to operate a municipal police force, make arrangements to receive policing from a neighbouring municipal police force, or contract with the RCMP for police service. British Columbia's provincial government is required to provide policing for communities smaller than 2,500 people at no cost to the community.

Apart from Quebec and Ontario, the provinces contract with the RCMP to provide regional police service. Ontario and Quebec have their own provincial police forces, the Ontario Provincial Police (OPP) and Sûreté du Québec (SQ). The SQ provides policing in most rural areas of Quebec. The OPP has five Rural and Agricultural Crime Teams (RACTs), formed in 1997. RACTs provide specialised regional services such as 'crime analysis, investigative leadership and expertise to reduce break and enters, recover stolen property and curtail agriculture and forestry crime in Ontario' (Ontario Provincial Police 2006). RACT also handles drug concerns like grow operations and theft of anhydrous ammonia for use in the production of methamphetamines, as well as agricultural and timber theft.

The First Nations Policing Policy (FNPP) aims to 'provide Aboriginal communities with policing arrangements that respect their cultures and ways of life' (Aboriginal Policing Directorate 2007; Solicitor General Canada 2004). First Nations policing in Canada takes a community policing approach, seeking to promote police services that are accountable to individual communities through self-administered police services and collaboration with the RCMP (RCMP 2008b). The structure of First Nations policing is established for individual communities and includes Self-Administered Agreements, First Nations Community Police Service Framework Agreements, and Community Tripartite Agreements (CTAs) (Solicitor General Canada 2004). As of 2007 there were '319 Aboriginal communities in Canada with dedicated police services employing close to 1,000 police officers, most of whom are of Aboriginal descent. Some of the services are self-administered, while others are managed through the RCMP' (Aboriginal Policing Directorate 2007).

Policing and Rural Society

Increasingly, local communities are required to bear the cost of policing under neo-liberal policies promoting regionalism. The offloading of federal responsibility for the cost of policing onto the provinces leads to passing the cost of policing on to municipalities even as local control of policing decreases due to regional administration of policing (Lithopoulos and Rigakos 2005). Between 1928 and 1966, the federal government paid 60% of the costs of policing in local communities. Between 1966 and 1990 that had dropped by half and continued to slowly decline (Solicitor General Canada 1996). As a result, local policing is increasingly contracted by the province or municipality to the lowest bidder, reinforcing local concerns about the adequacy of policing in individual communities. These concerns are greatest for rural areas which fear decreased visibility of officers and decreased autonomy under regionalisation. Critics have noted that despite the justification of regionalisation as cost effective and efficient, Canadian policing moved to community policing models alongside the implementation of regional management, decentralising policing functions in the face of centralising organisational changes (Lithopoulos and Rigakos 2005).

Although the rural Canadian population has generally grown over recent years, there has been a dramatic decline in the availability of local police services in rural and small town Canada. Aggregated data reveals a 26 percent decrease in local police services from 1998 to 2005. In fact, most rural and small towns are now serviced by police agencies located within a 30 minute drive (Halseth and Ryser 2006). This makes it less likely that crime in rural places, and especially domestic crimes will be reported or policed (see also Squire and Gill, this volume). This is exacerbated by geographic and social isolation, such as from social services or victim support (DeKeseredy *et al.* 2009), and the absence of public transportation (Lewis 2003).

Rural Crime and Policing in the USA

Cultural Background

Most US citizens also abide by the stereotype that their rural communities must be crime free. Yet in a comparative study, Donnermeyer (2007b) noted that violent crime rates in some of the USA's most rural counties exceeded that of 51 metropolitan areas: 'aggregated to a national level, urban rates may be higher, but there are plenty of exceptions to that generalisation, and these multiple exceptions call into question ... the myth that rural crime is unimportant' (Donnermeyer 2007b: 7). There is also growing evidence that domestic crimes and abuse are significant in rural USA. A recent study by Haight *et al.* (2005) highlights the physical and mental abuse suffered by children of methamphetamine abusers in a rural Midwestern county. Research by DeKeseredy and associates (DeKeseredy *et al.* 2007; DeKeseredy and Joseph 2006), Gagne (1992), and Websdale (1995a, 1995b, 1998) graphically illustrates the extent of violence against women, especially among married and co-habiting couples in the rural USA. Further, these studies demonstrate the extent to which rural community norms tolerate and even encourage this kind of violence. They demonstrate that it is social organisation, not disorganisation that represents the broad social structural and economic conditions behind rural crime. Yet, not only are the rural police confronted with crimes that are sometimes hidden away by norms of tolerance or the invisible underclass within rural communities in the USA (Smith 2008), they are expected to express a style of enforcement that is different from big city police (Weisheit *et al.* 2006).

The Structure of Rural Policing

The structure of policing in the USA is strikingly different from Canada, even though their rural sectors may share many of the same issues and challenges in relation to crime. It is less centralised than Canada's, being based on the concept of federalism or shared power between national, state and local governments. The result is a mind-boggling complex system of policing that is simultaneously centralised and decentralised, with overlapping jurisdictions and legal mandates for investigatory functions, enforcement and arrests.

In the USA, national level agencies, such as Alcohol, Tobacco and Firearms, the Federal Bureau of Investigation, Federal Marshals and the Drug Enforcement Administration, have jurisdiction over certain crimes in rural (and urban) areas that involve criminal offenses as defined by Federal law. In turn, at the state level are state police/highway patrol and numerous enforcement agencies associated with various departments that possess police powers to investigate violations and make arrests. Then, there are an incredible number of local law enforcement agencies, ranging in size from thousands of sworn and non-sworn personnel in big city departments to those with only a few part-time employees, representing the smallest and mostly rural located agencies. These local agencies consist of county police and sheriff's departments, township law enforcement, and police agencies for villages, towns and cities of various sizes. Local law enforcement agencies exercise a large degree of autonomy in their operations, even though they are subject to state and federal laws/regulations for legal definitions of crime, minimum standards of training, and specifications for patrol cars, communication equipment, and other police paraphernalia. Added to this already complex and segmented system are the various special duty task forces consisting of police officers from multiple jurisdictions assigned to special crimes, such as human trafficking and drug production. Finally, special jurisdictions, consisting of local, state and national park police, plus agencies associated with tribally operated law enforcement agencies, round out the menagerie of policing in the USA.

According to the Bureau of Justice Statistics, there are nearly 18,000 separate local, state and special law enforcement jurisdictions in the USA (Bureau of Justice Statistics, on-line: www.albany.edu/sourcebook 12/22/08). Of these, over 70% are local and another 17% are sheriff's department with county-wide responsibility (there are about 3,200 county or county equivalents in the USA). In particular, the concept of the sheriff historically comes out of the British system of an appointed protector of the King's property, called a "shire-reeve" (Falcone and Wells 1995). Today, however, the sheriff is an elected office, hence, every sheriff must deal with the political realities and dilemmas of enforcement within the context of re-election after a typical four-year term.

Local law enforcement agencies employ over one million individuals, of whom about 75% are sworn personnel with powers of arrest. Although there are no truly definitive numbers, it would appear from various statistical databases, including the 2004 Census of State and Local Law Enforcement Agencies (US Department of Justice, Bureau of Justice Statistics 2007), that about 250,000 police officers are members of 12,000 police agencies whose jurisdictions are significantly or totally rural.

To illustrate one problem of rural law enforcement in the USA, the proportion of personnel in a police agency who are sworn with arrest powers increases dramatically as the population size of the service area declines. In agencies with service areas of less than 10,000 in population (most of which, but not all, are rural), nearly 90% are sworn. In large, urban areas, the proportion is below 50%, indicating that the big agencies have many more support staff and specialists, from forensics to clerical, to realise their mission. In contrast, the smallest rural police agencies in the USA are composed of only a few personnel, all of whom are sworn and many of whom are part-time officers (US Department of Justice, Bureau of Justice Statistics 2006). As well, the operating budget declines from the larger to the smaller agencies, as does the base salary of officers (US Department of Justice, Bureau of Justice Statistics 2006). For example, in sheriff's departments serving areas with 100,000 or more persons, the average operating budget per sworn officer is about $130,000 (USD), compared to about $82,000 for those serving 50,000 or fewer persons. Entry-level base salaries decline from $38,800 (USD) for sheriff's offices serving one million plus residents, to $23,300 for those serving populations of fewer than 10,000 persons (US Department of Justice, Bureau of Justice Statistics 2006). As Falcone *et al.* (2002) and Maguire *et al.* (1991) note, rural law enforcement agencies in the USA are more generalist oriented and low-tech in character than their larger, more complex and better resourced urban counterparts.

Reservation land is for Native Americans in the USA through treaties negotiated as the Europeanisation of the country spread inexorably west. Although treaty violations form a long, sad chapter in the history of the USA, many reservation areas have gradually come to operate as sovereign and autonomous nations. This includes the formation of professional police forces. By far, the largest reservation police force is the Navajo National Department of Law Enforcement with over 320 sworn personnel who are responsible for an area that extends over four states of the southwest. However, many other reservation lands have police agencies with several dozen officers each (US Department of Justice, Bureau of Justice Statistics

2003). These agencies have a daunting job. Decades of treaty violations and exploitation of resources on reservation land, discrimination by the majority white population, persistent and deep-seated poverty, and the emergence of very high rates of suicide, alcoholism, interpersonal violence for some Native American peoples, creates a difficult environment for many of these police agencies (Weisheit *et al.* 2006). As with other local law enforcement in the complex web of the US system, these Native American agencies engage in cooperative investigatory, enforcement and arrest activities with state and federal law enforcement departments, and all with varying degrees of inter-agency cooperation and competition.

Policing and Rural Society

Jiao (2001) found that rural residents prefer that rural police adopt a problem-solving approach rather than approaches based on a model of police professionalism. Rural residents expect their law enforcement to be involved in a wider array of community service and public safety related functions than a traditional enforcement mode (Wood and Trostle 1997). A review of police activities in one rural community found a variety of crime and community related calls for service (Payne *et al.* 2005). On the one hand, they must respond to calls about assault (both in public places, like bars, and in private places, like homes), vandalism and theft. On the other hand, they must also respond to calls for such mundane things like loose dogs or youth overheard using obscene language.

Residents of rural US communities value forms of insider status as much as competence in traditional police functions among the officers who serve them (Falcone *et al.* 2002). This takes various forms. Officers are insiders if they grew up in a rural community or on a farm. If officers are serving in the same community where their parents or other members of the extended family live, or if they graduated from the local or a nearby high school, then their insider status is even more enhanced. Residents assume they understand local norms and values, and have a greater appreciation of community strengths, and a greater willingness to deal appropriately with calls for service that maintain public order and selectively enforce laws (Falcone *et al.* 2002; Payne *et al.* 2005). This level of familiarity also creates conflicts of interests among officers who must consider arresting violators whom they know, or whose family they know. It also means that they may ignore situations in which violations are committed by citizens who form the small, inner circle of local elites and influentials who frequently dominate the political, economic and cultural power in small town America. For these reasons, violence against women and other crimes may not be strenuously enforced, or enforcement may be selective (DeKeseredy *et al.* 2009; Weisheit *et al.* 2006).

Under the shared power or federalism model of the US, small towns and other local political entities may have either a 'strong mayor' who runs the community, or a council of elected officials who hire a 'manager' to administer local government affairs. In either form of governance, or in the case of sheriffs who hold elected office, rural officers who enforce laws and otherwise conduct themselves in ways that are in conflict with community norms or the vested interest of local elites, may soon find themselves unemployed (Weisheit *et al.* 2006). As well, because of the better pay and retirement, health insurance and other fringe benefits offered by better resourced agencies serving larger populations, turnover can be a problem in rural communities (Wood 2002). With the emergence of organised crime (both marijuana and methamphetamine production, as well as theft rings related to livestock and farm equipment) (Mears *et al.* 2007a; Weisheit 2008), gangs (Donnermeyer 1994; Weisheit *et al.* 2006), human trafficking (Wilson and Dalton 2007) and other crimes which were much less frequent in the past, rural law enforcement in the USA finds itself in a tough situation. Community expectations for the rural police are embedded in a heritage of preferences for a more generalist approach that is in part reflective of discretionary behaviours compatible with localised expressions of vested interests, inequality, and uneven distribution of power and decision-making about peoples, networks and groups.

A New Realism for Rural Canada and the USA

Rural police in both Canada and the USA operate in a challenging context. Their jurisdictions are quite large, and they must enforce laws, investigate violations and make arrests in the context of a complex, segmented and overlapping system of police agencies whose missions do not completely align and who often compete for resources and recognition/credit for performance. The smaller the agency, the fewer resources are available and the less able is the agency to compete in an open job market for the best officers. Canada tackles these problems through the RCMP and provincial police of Ontario and Quebec, but even so, the rural sectors of both countries receive less police attention, resource-wise, than the big cities.

We conclude this chapter with a discussion about future research on policing in rural Canada and the USA. We contend that the present literature on policing in the rural context of both countries is inadequate because policing scholars largely ignore criminological theory in general, and rarely adopt critical theory (Young 2002). With a more critical approach, realities long hidden can be revealed, and although they may seem new, they are nothing more than persisting features of rural places that show how the social, economic and normative structures of rural communities enable crime and frame the ways in which rural police do their jobs.

DeKeseredy *et al.* (2006) argue that it is more than demographic characteristics, social structural factors, and economic indicators that predict crime. Quantitatively and qualitatively derived measures of the social and economic dimensions of places, both rural and urban, are mere indicators of realities that help us to understand the relationship between economic decay, racial/ethnic and other forms of segmentation, and crime. It is these realities that help us to understand the relationship of policing to crime by critically examining the location of law enforcement within the social and economic structures of inequality.

Hence, framing a discussion of policing rural Canada and the USA requires stepping back and considering the rural realities of smaller towns and agricultural areas. Aside from the immediate issues of policing large, rural tracts with comparatively fewer people than in the city, there are the issues of poverty, violence, organised drug production, the emergence of gangs, and native or 'First Nations' peoples of both countries who have long suffered from the presence of Europeans on or nearby their lands, as is true of countries like Australia (Hogg and Carrington 2006).

At the present time, research on policing rural Canada and the USA is mostly descriptive and atheoretical (Weisheit *et al.* 2006). Not only is there is dearth of comparative studies across differing kinds of rural communities in either country, there are no studies which have attempted to link the interplay of policing within the context of rural social structure and culture. Simply put, the studies are either of single locations or empirically based surveys of multiple rural agencies on such mundane criminological phenomena as policing styles (i.e. traditional enforcement versus community oriented policing or the more generalist nature of rural police), turnover of rural sworn personnel, and calls for police services by rural citizens.

Unfortunately, what is left out are the ways rural policing styles are embedded in social and economic structures that define the distinctive features of enforcement in rural Canada and the USA. Certainly, given the diversity of rural communities and people, variations in the ways police differentially enforce laws and conduct themselves with citizens (both criminals and law-abiding members) would provide valuable insights for criminological theories of policing, and for the development of appropriate strategies. One way to blaze a new trail for the criminological study of policing in rural Canada and the USA is to adopt a definition of place similar to Liepins' (2000). To quote (Liepins 2000: 29): 'an approach to community must determine a way in which the population involved can be treated as a set of heterogeneous figures who constantly locate themselves in multiple positions and groups.' Hence, communities are 'temporally and locationally specific terrains of power and discourse' (Liepins 2000: 29). Within these rural communities are various kinds of social orders that interplay with the formal structures of policing in rural Canada and the USA to define responses to differing kinds of offences, from intimate partner violence to agricultural crime.

Without connecting the specifics and the mundane to these larger structures and processes, understanding policing in the rural context of Canada and the USA will continue down a path of testing specific actions but without ever considering macro-level forces that place strong parameters for the ways in which rural communities, peoples and police agencies can change and work together in more effective partnerships.

There is now growing evidence that Canadian and US urban policing practices are becoming more alike (DeKeseredy *et al.* 2009; Grabb and Curtis 2005). For example, during the late 1990s, the Ontario provincial government followed the advice of US criminologist George Kelling and that of former New York City Police Commissioner William Bratton and implemented aggressive policing strategies aimed at panhandlers, public drunkenness, and other types of social or physical disorder (DeKeseredy *et al.* 2003). Yet there is a growing body of literature that raises serious doubt that social disorganisation causes crime and collective efficacy reduces it (Donnermeyer and DeKeseredy 2008). Rural realities simply point to the idea that social organisation simultaneously enables and constrains different forms of what both countries, through their differing styles of governance (including policing styles) and system of laws, define as criminal behaviours (Donnermeyer 2007a). Although, to the best of our knowledge, there are no published comparative studies of Canadian and US rural policing, it is logical to assume that there are many other similarities. It may, yet again, be painfully obvious, but worth stating anyway: more research on the similarities and differences between Canadian and US rural policing is necessary.

Chapter 4

Policing the Outback:
Impacts of Isolation and Integration in an
Australian Context

Elaine Barclay, John Scott and Joseph F. Donnermeyer

It's more about community here, you get to finish a job properly which is very different to working in the city where you may arrest an offender, lock him up and then lose track of him because he is now in the bigger system. Here we know the offenders and their families and we constantly participate in community activities so we soon become part of the community ourselves.

(Newsbeat 2007a)

Integration is important in narratives of policing in rural Australia. Communal integration, often depicted in rural studies as a positive element of gemeinschaft relations, or more recently equated with social capital, is crucial in understanding how remote communities are policed in Australia. However, while the above quote presents integration as a positive achievement, we wish to highlight the complexity of social relations in rural Australian settings and how this has differing implications for how police work is conducted. In particular, we will examine how particular visions of social order are achieved and maintained through practices of 'boundary maintenance', involving the material and symbolic inclusion and exclusion of specific individuals and populations. As such, rural spaces are not to be conceived as homogenous entities, but rather are diverse and pluralistic settings with competing and hierarchised normative communities.

This chapter explores these issues with reference to the experience of policing in remote Australian communities. We are concerned with how the material conditions of rural police work and symbolic understandings about the impact of 'rurality' upon police practice. In doing so, we acknowledge the links between space and policing, noting the spatial influences on the normative frameworks which guide police work. We wish to examine how integration in rural contexts implies adherence to normative frameworks, sustained by practical and symbolic policing measures, which operate to include and exclude specific populations marked as 'troublesome'. In particular, we will discuss how policing operates to reinforce normative accounts of Aboriginality and materially subjugate and exclude Indigenous populations.

Idyllic Policing

To understand crime we must look at the social reaction to it, including the meaning ascribed to it. Crime and crime control are enmeshed in the wider moral-politics of communities, forms of cultural mentality and sensibility and larger strategies and rationales of power that comprise forms of rule. As Hogg and Carrington (2006: 62) have argued:

> the imaginative constructs of the rural community have powerful effects in policing the micro-social spaces of rural family life rendering the violence and other dysfunctions in it relatively hidden from public view. As a social and cultural construction, the imagined rural community operates as a device to include and exclude, silence and deny, and to render certain aspects invisible and visible.

Policing may act to produce and maintain both the material and symbolic boundaries which constitute a hegemonic, but contested, sense of the rural idyll (Bell 2006; Erikson 1962; Yarwood 2007b and this volume). Popular articulations of rurality in Australia have had a productivist and masculinist character (see Carrington and Scott 2008): images of farmers, stockmen and miners are integral to our understanding of the Australian 'bush' and 'outback'. Outside of cities, Aborigines are placed in an exotic context for consumption, while cultivated and productivist rural landscapes are peopled by farmers (Hogg and Carrington 2006: 4). Such views tend to obscure aspects of difference and division in the countryside, producing a specific socio-cultural articulation of gemeinschaft.

Indigenous Australians have different conceptions of space, which regularly clash with these constructs. Open air spaces within towns, rather than being sites for tourist consumption, may be considered as places to congregate and drink. Aborigines, as such, are defined as disrupting the Non-Indigenous social order and are marked as nuisances and defined as dirty and untidy. To borrow a phrase from Mary Douglas (1992), they may be considered as 'matter out of place'. The irony of Aboriginals being misplaced in the Australian landscape should not be lost. This construct in turn results in high levels of policing, non-comparable with Non-Indigenous populations, and high arrest rates for minor offences, such as bad language (Yarwood 2007b).

'Social problems' are dramatised as threats to culturally specific characterisations of rural life. Many rural communities have experienced loss of services, depopulation, economic decline, and erosion of infrastructure (Pennings 1999), while others have experienced population growth, economic boom or re-invention as tourist destinations. These changes have threatened to disrupt the established social order of many communities. Attempts to maintain social order are reflected in official and popular discourses which present a picture of increasing crime rates in rural Australia (Pennings 1999) that have, in turn, amplified fear of crime in rural communities and, in some instances, generated an excessive policing response to perceived problems (see Hogg and Carrington 2006; Lee 2007).

Before examining how this 'architecture of rural life' (Hogg and Carrington 2006) might shape policing practices, leading to the over-policing and under-policing of specific individuals and populations, an account of the structural conditions in which policing is conducted in rural Australian settings is required.

Context

Australia has a land area of almost 7.7 million square kilometres, yet the majority of Australia's 21 million residents live within two coastal regions. The remaining population is scattered across a vast region commonly referred to as 'the outback' that varies from tropical regions in the North, to flat, arid and sparsely vegetated expanses in the interior, to more temperate zones in the South (Australia Bureau of Statistics 2006). It is difficult to be precise as to where Australia's outback actually begins, yet, as one travels inland, farms become larger, towns become smaller and further apart, and at some indefinable point, these areas become *the outback*. Communities and crime rates are extremely varied (Barclay *et al.* 2007; Jobes *et al.* 2000), ranging from relatively low in bucolic 'country' communities, to very high in a few 'outback' towns.

The organisation of policing in Australia is quite distinct from both the USA and UK. Training statutes, laws and organisational practices are more tightly controlled in Australian states and territories, each of which is policed by a single organisation. Currently Australia has eight police services, centrally controlled and organised within each state and territory. Each state police service is divided into regions, which are subdivided into police districts or Local Area Commands. Each command has various police stations, with some small rural stations staffed by only one officer (Adcock 2002; NSW Police 2008). Many small villages are serviced by officers located in neighbouring towns, with officers working between communities as demand requires (Jobes *et al.* 2000).

Physical isolation impacts upon all aspects of rural policing (Moore 1992). For example, police services in all states of Australia experience difficulty in the recruitment and retention of officers to rural areas. Some of the more remote regions, which do not conform to representations of the gemeinschaft, are not popular postings (Adcock 2002). The cost of living, lack of high quality and affordable housing, the lifestyle, educational access for children, and the lack of employment options for partners are cited by official policing publications as the main deterrents (Newsbeat 2007b). However, such locations also have histories of racial conflict, the public face of which has been grounded in an apparent mutual hostility between police and Indigenous communities. Racial and ethnic discrimination have been reported to be common in many rural areas (Coorey 1990; Cunneen 1992).

A remote placement may be culturally and professionally challenging, particularly for officers who come from cities or large regional towns. Officers need a range of skills to successfully manage small and single unit police stations, such as a lockup keeper, manager of legal briefs and rural crime investigator (NSW Police 2007). Adcock (2002) believes that rural policing should be acknowledged to be a specialist role which requires specific and unique training, with incentives to attract officers to rural posts, such as higher salaries, opportunities for tenure and promotion and high standards of housing (Adcock 2002). In all states, tenure policies in rural communities have fluctuated between preferences for long terms to better integrate officers into rural communities to a preference for shorter terms to facilitate promotion and staff experience. Short terms of two to four years are seen to be preferable (Adcock 2002) to *prevent* complete integration into the community, which is believed to hinder officers' effectiveness in administering the law in places where they know almost everyone.

Physical isolation also impacts upon how police do their work. For example, rural residents become concerned when telephone calls to their local station are diverted to the nearest regional centre if the station is unattended. When regional centres are more than two hours drive away, police dispatched to emergency situations cannot provide a timely response and have limited knowledge of the area, the location of properties, and road conditions (Jobes *et al.* 2000). Many remote communities are not serviced by all weather roads. The closure of roads and airports in wet conditions can leave some communities isolated for days and even weeks (Jobes 2003). In small stations, being in charge of one's own activities is an attractive part of rural policing, even though officers are answerable to a supervisor who is some distance away (Adcock 2007). Although most officers try to keep their station open, they can close it if needed to visit local farms, conduct patrols, make enquiries or talk to local residents (Bayley 1986).

Remoteness also impacts on access to centralised support services. While improvements in communications systems have enhanced policing capabilities in remote areas, distance can still impact upon decisions to supply centralised support functions to major incidences or investigations. Issues of staff time, travel costs and the need to maximise scarce resources are considered when making decisions regarding provisions for remote areas (Barclay and Donnermeyer 2007; Pennings 1999).

The issues of remoteness and resourcing, cited above, mean that law enforcement is only one function of policing, particularly in rural settings (Swanson, Territo and Taylor 1988). Rural officers tend to be generalist police who handle all varieties of crime, often spending more time in service roles than law enforcement. Rural police in Australia deal with the same types of crimes experienced in urban centres, such as traffic offences, drug offences, alcohol violations, disorderly conduct, anti-social behaviour, vandalism, assault, domestic violence, break and enter, and theft (Jobes 2003). Police also deal with rural-specific crimes such as livestock theft and other theft from farms, large scale cannabis production, illegal fishing or hunting, and other environmental crimes (Pennings 1999). These crimes often require specialised responses that may involve a number of external agencies. While the states of Queensland, New South Wales, Tasmania and Victoria have specialised rural crime investigators, in most situations it is the local police officer who is required to mobilise and coordinate collective multi-agency and multi-jurisdictional responses, which demands broad generalist capabilities (Pennings 1999). In addition to enforcing laws, officers are often called upon to assist in medical emergencies, conduct community education, crime prevention and youth programs, and manage special events (Jobes 2002). Commensurate with community expectations, officers are as likely to administer to tragedies and accidents as they are to 'serious' crime. These 'welfare' or 'humanitarian' duties establish officers as good citizens and indirectly facilitate law enforcement and reduce crime. Given the emphasis on generalist service provision, what develops is a 'localistic', as opposed to 'legalistic', approach to policing.

Localistic Policing

O'Connor and Gray's (1989) study of Walcha, a small and relatively isolated township situated in North West New South Wales, found police possessed a positive attitude to their work and the town they policed. They perceived Walcha as a low crime area and believed the public supported them. While the community sought to restrict the police in some areas, residents supported the maintenance of order around norms of sobriety, protection of property and person, and respect for authority. Walcha's inhabitants considered their community to be virtually crime free, despite the crime rate being in line with the national average. The social isolation and socio-demographic profile of Walcha revealed a community that was largely culturally homogeneous, characterised by strong intergenerational and horizontal ties to locale. The tight-knit nature of the community meant that when crimes did occur they were not perceived as socially threatening. As such local police only occasionally felt required to initiate formal criminal action. When anxiety regarding crime did manifest, it was bound up with suspicion of the unfamiliar. It was associated with outsiders and events which brought outsiders into the community, disrupting localised narratives of the idyll. The externalisation of danger correlated with the stratification of life within the community, with length of residence informing social position, status and authority.

Similarly, in a more wide ranging study of rural townships, Jobes (2002, 2003) found that most police units working outside of highly fragmented communities enjoy their work, and take advantage of the rural setting and lifestyle. Many police said their lives had improved since being assigned to the country and few wanted to go back to the city. However, Jobes also cautioned that officers in racially divided towns with high crime rates spent considerable time enforcing the law to an Indigenous minority whom they regarded as antagonistic and uncooperative. Similarly, Cowlishaw (1987: 43) argued that racial dichotomisation in 'Brindleton' led to conflict 'which is endemic in many New South Wales country towns'. Officers endured Brindletown solely to be promoted in the state system. Police avoided making waves in the general community while enforcing the law against a hostile and repressed Indigenous population. Other authors have examined specific types of rural law enforcement problems. It is important to note here, however, that outwardly culturally homogenous communities also have specific moral-politics which may influence police work. With regard to domestic violence, Coorey (1990) reported that police in 'Barken' categorised women as 'good' or 'bad'. Good women were rarely charged and mateship in the town reduced the ability of the police to objectively enforce the law.

Domestic violence is extremely common in rural areas and is also significantly unreported (Hogg and Carrington 2006). Social and logistical obstacles to reporting in many country towns are likely to be far greater than those facing urban residents (Coorey 1990). Accessing services for victims of domestic violence is a problem when no refuge or public transport is available in remote communities (Carrington 2007). Research indicates that police have been reluctant to consider domestic violence as a law and order issue in rural settings (Hogg and Carrington 2006). Research has consistently shown how police in rural communities are less likely to initiate formal actions against perceived private infractions (see, for example, O'Connor and Gray 1989). Greater importance is accorded to establishing and maintaining public tranquillity or 'the peace', as opposed to imposing law and order at any cost (see Jobes 2002). Hogg and Carrington (2006: 154) have observed that while rural crime investigators are afforded special significance in police cultures and many rural communities, the work of domestic violence officers is less likely to be valued. While they found the work of such officers has a significant impact on the problem of domestic violence and that these officers were professional and committed to their work, the officers expressed frustration at lack of community recognition and support for their work. The material conditions of rural policing, described elsewhere in this chapter, also presented difficulties for police attempting to address the problem of domestic violence. For example, many small stations are unmanned at night, resulting in a lack of availability of police when this criminal activitiy was most likely to occur (Hogg and Carrington 2006: 154).

Police culture strongly influences how police think and act (Bouza 1990; Manning 1977; Young 1993) and there is much evidence to indicate that police culture, while present in rural settings, is quite distinct from urban manifestations, producing distinct issues to those encountered in urban settings (Jobes 2002). The material conditions of policing in geographically remote and isolated communities, influences how policing is carried out (Bayley 1986). For example, Australian police typically reside within rural communities; their private lives being closely intertwined with their public role. Successful officers are likely to be integrated into a local community and make effective use of established local social networks, adopting what might be considered a community based model of policing. While this model can produce successful results, with integration into informal social networks providing police increased opportunities to solve crime, rural police regularly find themselves occupying competing roles of law enforcer and local citizen. Moreover, when officers first arrive in a community, they establish through discussion and action what behaviour will and will not be tolerated. This normative process is a product of interaction with the community and integration into certain social networks within the community (Bayley 1986).

With a localistic model of policing, the public influences how policing is carried out, defining important aspects of the role of the police, including police discretion. In contrast to metropolitan policing, where discretion is largely influenced by the organisational structure of a department, rural police discretion is likely to be influenced by the way in which the community is organised. More significantly, officers are guided by their local knowledge of residents, and expectations of the community (Adcock 2002). This influences discretionary practices, resulting in the under-policing and over-policing of certain groups and/or activities. In a study of farm crime Barclay and Donnermeyer (2007) found residents feared being accused of informing on a person to the police. The identification of a neighbour or someone else in the community as a possible offender can have serious ramifications for the victim, the accused, their families and friends, and the whole community. Officers interviewed were well aware of the reasons why farmers fail to report crimes, and the pervasiveness of a rural ideology of self-reliance within the culture (Barclay and Donnermeyer 2007).

While localistic policing can be viewed largely as a positive development, immersion into the community can exacerbate the problematic aspects of police work. Local officers are expected to become part of the community even if it compromises objectivity. A legalistic style, while limiting opportunities for integration and communication, might be less sympathetic to privileged groups in the community, avoiding police bias and discrimination. While rural communities tend to have a strong sense of geographic and social identity, the problem remains: who is defined as *belonging* to the community. Residence alone does not signify belonging. For example, in many rural Australian communities, to be valued, an individual or group must be regarded as contributing to the general prosperity of the community, signified in terms of 'productivity'. Jobes (2003: 11–12) found that police in rural NSW identified 'types of people who disproportionately accounted for law enforcement problems'. These included seasonal workers, people in transit and visitors. Youth, excessive drinkers and Aboriginals were also mentioned as problem populations, highlighting how the spatial politics of rural communities may mark out specific groups or activities as a threat to social stability.

Policing Indigenous Populations

In contemporary Australia, Aboriginal people are over-represented both as victims and as perpetrators of all forms of violent crimes (Carrington 2007; Cunneen 2005; Fitzgerald and Weatherburn 2001; Harding *et al.* 1998). While Aboriginal Australians constitute approximately 2% of the population, they comprise about 20% of all prisoners nationally (Australian Bureau of Statistics 2006; Blagg and Valuri 2004). The reasons for Indigenous people's over-representation, both as offenders and victims, is multifaceted. Cunneen (2006) cites specific factors necessary to explain Indigenous over-representation, which include:

- The impact of policing (in particular, the adverse use of police discretion and availability of diversionary options).
- Legislation (especially the impact of laws giving rise to indirect discrimination, such as legislation governing public places or alcohol).
- Factors in judicial decision-making (in particular, the availability of non-custodial options).
- Environmental and locational factors (especially the social and economic effects of living in small rural and remote communities).
- Cultural difference (such as, the use of Aboriginal English).
- Socio-economic factors (in particular, high levels of unemployment, poverty, lower educational attainment, poor housing, poor health).
- Historical legacies, including the impact of specific colonial policies (especially the forced removal of Indigenous children).

Police resources have been disproportionately allocated to racially segmented towns in the same way that they have been in urban racial and ethnic enclaves. In such places, police often distinguish between 'good' and 'bad' Aboriginals. 'Good' Aboriginals are usually well behaved, recognising and adapting to a subordinate role in the local social hierarchy. 'Bad' Aborigines are perceived to be defiant or disrespectful, refusing to assimilate into the dominant normative culture (O'Connor and Gray 1989). Many Aboriginals have evolved an 'oppositional culture' sensitive to discrimination and hostile to the dominant culture and law enforcement. Police respond in a similar fashion, stereotyping Aborigines as violent, uncooperative and disorderly. Violence in such towns is presented as an Aboriginal problem by civic and community leaders. Each group sees themselves as victims, the police casting themselves as having a difficult job with little support. Extreme polarisation is the result. To avoid disrupting the dominant normative system some officers practice reactive policing, which may involve prejudice, abuse of powers and cultural insensitivity.

Aboriginals are rarely portrayed as victims of crime. Nevertheless, Indigenous women and children suffer significantly higher rates of interpersonal violence and abuse compared to Non-Indigenous women. Patterns of victimisation are likely to be concentrated in particular areas where poverty is highest and the opportunity to access services is lowest (Cunneen 2006). Family violence accounts for 63% of all Indigenous homicides compared with 33% of non-Indigenous homicides. Aboriginal women are also less likely to report crime than non-Indigenous women (Aboriginal and Torres Strait Islander Women's Task Force on Violence 2000).

A reason for the lack of attention to Indigenous people as crime victims is a focus in rural communities on stranger assault, property crime and drug and alcohol abuse, behaviours that people associate with 'the crime problem' (Lee 2007). Moreover, residents are concerned with crimes which occur in the central business district or parks of country towns, especially outside of regular working hours. It is hardly surprising that Indigenous people tend to use these spaces for recreational purposes and tend to be restricted, through lack of mobility, to these places after hours. On the other hand, interpersonal violence, especially in domestic space, tends to be under-reported in rural settings (Hogg and Carrington 2006).

While family violence is a very real and significant problem in Indigenous communities, such violence is represented as a uniquely Indigenous problem. In many rural towns, Indigenous communities comprise a distinct and visible cultural entity. Because of their living conditions, many Indigenous people are unable to maintain domestic privacy. Indigenous life is more publicly visible and subject to adverse cultural judgements and formal interventions by the police as violations of public order. Indigenous life is exposed to adverse cultural judgment and formal intervention as a violation of civic decency. This results in the visibility of Indigenous violence, while non-Indigenous family violence is masked from public view. Crucial to this scenario is the idea that stronger community bonds in rural Australia mitigate against violence (Hogg and Carrington 2006: 136). In stark contrast to this, Indigenous private life has been defined as a cause of crime problems, rather than a solution.

This scenario was evident in the response to a 2007 report into child abuse within Indigenous communities in the Northern Territory. The report documented high levels of drinking, drug use, sexual abuse and lawlessness in Northern Territory Aboriginal communities. The Howard government characterised the situation as a national emergency, which required a unilateral response targeting lawless Indigenous communities. This involved placing police within communities, placing bans on alcohol, drugs and pornography, providing health checks for Aboriginal children under the age of 16 years, and quarantining 50% of welfare payments for the purchase of food and other essentials. By mid-2008, more than 8,000 children had received health checks and there were 51 extra police stationed in remote communities (Edwards 2008).

New Policing Strategies

Following the observations of David Garland, the Northern Territory intervention might be considered a reversion by the state to punitive strategies which have their roots in criminologies of the 'alien other'. These represent criminals as dangerous members of distinct racial and social groups which bear little resemblance to 'us' (Garland 1996: 461). However, the Northern Territory intervention at first sits odd in a period in which governments have withdrawn or at least qualified their claims of sovereign control. New strategies have appeared in late modernity described as 'new criminologies of everyday life' which hold the premise that crime is a normal aspect of society and a risk to be calculated. Moreover, as these new strategies argue, crime is not to be addressed by state agencies, such as the police, but by the organisations, individuals and institutions of civil society. This criminology employs responsibilisation strategies so that the state acts indirectly to activate action on the part of non-state agencies and organisations. The recurring message of the approach is that the state cannot act unilaterally to control crime, but must work in partnership with civil society (Garland 1996).

In light of such developments, the Northern Territory intervention might be considered a symbolic denial of the normalised nature of crime and the inability of the sovereign state to provide security, law and order and crime control within its sovereign boundaries. It is a 'hysterical' reassertion of the old myth of the sovereign state (Garland 1996: 449). Yet the Northern Territory intervention is not typical of responses to perceived law and order problems in predominantly Indigenous communities. Policing of Indigenous lands has been under resourced, with most communities lacking full time police services and under-staffed (Yarwood 2007b). Nonetheless, the idea of a 'rural crisis' has placed pressure on policy makers to work with communities to address ongoing social problems.

This shift with regard to Indigenous policies can be traced back to the Royal Commission into Aboriginal Deaths in Custody established in 1989. Since this inquiry, police agencies in partnership with their state governments have made significant changes to reduce the disadvantage experienced by Indigenous people within the criminal justice system. New programmes include the adoption of community policing, diversionary programs, cross cultural training and education, and greater Indigenous autonomy with regard to justice issues. Key strategies have included:

- *Aboriginal Community Patrols or Night Patrols* which are community based services that operate a safe transport and outreach service for people who are on the streets late at night. The patrols aim to reduce the risk of people becoming involved in crime and anti-social behaviour, either as potential victims or offenders, and reduce contact between Aboriginal people and the police. Patrols also provide other services, such as mediation in disputes, maintaining the peace at sporting events and acting as a nexus between police, courts, clinics and family (Blagg and Valuri 2004).
- *Circle Sentencing* is an alternative sentencing court for adult Aboriginal offenders. It involves taking a sentencing court to the local Indigenous community, where the magistrate, the offender and victim and their families and respected members of the local community sit in a circle, discuss the matter and arrive at an appropriate sentence (NSWCPD 2008).

- *Aboriginal Community Justice Groups* are local groups of Aboriginal people who come together to develop ways to address local law and justice issues (NSWCPD 2008).

Cunneen (2007) observes Indigenous responses have also included holistic anti-violence programmes, community justice groups and Indigenous courts. Although many of these initiatives reside within the existing non-Indigenous criminal justice system, they can help us to understand how Indigenous justice might develop as a distinctive response to Indigenous crime issues. These initiatives, many of which have developed in rural and remote areas, can increase community self-esteem and empowerment, enhancing the development of effective and culturally informed governance structures for communities. These strategies demonstrate how remoteness from state centred sovereign powers can be an incentive to develop criminal justice.

Conclusion

There is an assumed orderliness of rural communities. Crime disrupts the stereotype of rural communities as cohesive, caring and wholesome. As noted previously, O'Connor and Gray's research in Walcha found that the tight stratification of the town led to an externalisation of threats – crime was related to outsiders. Moreover, concern about crime in Walcha reflected concern about unwanted social change, which was perceived as a threat to the idealised version of the community. Crime, as such, may present as a threat to the symbolic boundaries of community. In Walcha race was not considered an issue in terms of policing the social order, with the town's population comprising low numbers of Indigenous people. However, research has repeatedly demonstrated that in Australian towns with high numbers of Indigenous people, race is an issue. With respect to this, crime outside the city is not so much spatialised as it is racialised (Scott *et al.* 2007).

The localistic model of policing does not necessarily translate to an idealised form of community policing. Officers may be encouraged to favour particular social groups and sectional interests. Community groups may significantly influence how police determine who is and is not subject to surveillance. The outcome may lead to over-policing or under-policing of specific interest communities and activities. While policing is a partnership for some members of the community, for others it is experienced as an imposition. Policing maintains an elaborate system of internal spatial and social boundaries and hierarchies. The reality of rural spaces as exclusivist challenges the idealisation of such spaces as natural, egalitarian and inclusive (Hogg and Carrington 2006: 87).

Traditionally, the focus of policing is on public rather than private space. Policing is about public offences, with severe crimes in domestic environments relatively ignored. One result of this is that Aborigines tend to be viewed as perpetrators, rather than victims of crime. Crime in rural communities tends to be considered the preserve of publicly visible populations, such as Aborigines, whose lack of resources confines their mobility and limits their access to legitimate recreational activities, resulting in increased visibility and surveillance. High arrest rates of Indigenous people relates to a clash of values about the use of space. The condition of Indigenous people does not elicit the type of community response that we see in cases of communities effected by 'natural' disasters. *They*, after all, are the social problem. Like other marginalised populations in rural communities, Aboriginal people do not conform to rural stereotypes which emphasise productivity. Instead, Indigenous people have been associated with consumption, being presented as an unproductive population who do not conform well with a rural mythology emphasising stoicism, independence and hard work. More alarming is the association of Aboriginal people with social disorder, which does not rest well with the myth of rural environments as consisting of simple, ordered, cohesive and homogenous communities.

Chapter 5

Rural Policing in France: The End of Genuine Community Policing

Christian Mouhanna

Introduction

Today, insecurity and crime are widely associated with urban life yet, historically, the countryside in Europe was often considered to be more dangerous due to the activities of armed brigands and de-mobilised mercenaries (Muchembled 2008). To counter this threat, the *Maréchaussée* (later the *gendarmerie*) was established during the middle ages in France as a judicial, militaristic force to protect rural areas (Luc 2005), while cities organised their own protection through civilian forces. Historically, the French state preserved this dualism because each force could be used to keep the other in check (Lignereux 2002). In 1941, urban police forces were integrated in a single national organisation but the division between two separate forces was retained. The *Gendarmerie Nationale*, a military force, is still in charge of security in rural areas, whereas the *Police Nationale*, a civilian force, is responsible for the cities and their suburbs.[1] The *gendarmes* live in barracks and cannot belong to unions while the *policiers*, like other civil servants, have greater personal freedoms.

Although *gendarmes* have a reputation for being rigid in the enforcement of law and order, they are in practice more flexible than *policiers*. Indeed, their public image is much better than the police and they undoubtedly have closer relationships with the citizens in their districts. It may be argued that the *gendarmes* have established a model for community policing, even if it was never officially recognised by the authorities. But this tradition is now threatened by new rules and new career paths. The *gendarmes* have adopted the dominant model of the French *Police Nationale*, prompting more efficient, target-based policing. Indeed, the French government is now arranging a merger between the two forces, under the authority of the Ministry of the Interior. This, coupled with a growing desire by

---

1   There are two types of *gendarmerie* in France: the *gendarmerie départementale*, is the one stressed here, while the second, the *gendarmerie mobile*, is a special force used to control riots and demonstrations. Some *gendarmes* do transfer from the latter to the former in the course of their career, but the culture of the two forces differs considerably, for reasons that are explained in this chapter.

gendarmes themselves to live more private lives, has important implications for the policing of rural France.

This chapter draws upon interviews with *gendarmes*, elected officials and local partners (Matelly and Mouhanna 2007; Mouhanna 1997, 2000, 2002a, 2002b) in order to evaluate these changes. Particular attention is paid to their impact on community policing and police-public relations. The first section examines how the *gendarmerie* functioned until the end of the 1990s before attention is turned to recent reform and change.

The *Gendarmerie Nationale*: The Real French Community Police Force

On a map, the *Police Nationale* districts looked like islands surrounded by *gendarmes*. Prior to 2000, each French *département* (i.e. county) was composed of one or more *Police* districts (cities of over 20,000 inhabitants and their suburbs) and an average of twenty *gendarmerie* districts in the countryside (Carraz and Hyest 1998). Although the latter controlled 95% of territory and 50% of population, this did not reflect the real balance of power between the two forces.

As a result of a modernisation programme, The *Police Nationale* was considered efficient, modern and capable of conducting criminal investigations. By contrast, the *gendarmes* seemed old-fashioned, less able to cope with new types of crimes and little more than an auxiliary for the 'real' police force. This image was offset by the relationship the *gendarmes* built with the citizens they served. Until the end of the 1990s, the French population had a better image of the *gendarmes*, despite their military status, than of the *Police Nationale*. In addition, the *gendarmes* themselves were conscious and proud of the social role they played in rural areas (Laffont and Meyer 1990; Mouhanna 2001). This interaction can be attributed to the way in which *gendarmes* were structured, namely that they lived in the districts they policed and had few resources.

Structure and Organisation

Although working within a highly centralised, hierarchical organisation, *gendarmes* were far from their superiors. Until the end of the 1990s, the French rural territory was covered by more than 3,600 *Brigades Territoriales* (BT) each with their own team. Each BT consisted of six to 20 *gendarmes*, but the vast majority of them had less than eight. Of course, the BT was placed under the responsibility of a commander, but this commander was a non-commissioned officer. The hierarchical superior, the captain, in charge of an average of 10 BT, was located elsewhere, often far away from the majority of the BT. He only had a good vision of the work of the BT in his own area. Even if he sent a great many messages by various means, including radio or internet, it was impossible for him to supervise all his subordinates. He actually visited each BT only once a month and could never be sure that reports concerning any one BT were reliable.

Moreover, he only spent two years in that position, after which he changed for another job, whereas many *gendarmes* worked in the same BT for eight years or more. In this system, the superiors were considered less capable than the grassroots teams, because they lacked knowledge about the local environment. In this respect, the French *gendarmerie* was closer to the 'street level bureaucrat' model (Lipsky 1980) than to the armed forces.

Things were very different in the *Police Nationale*. Each police district was (and still is) a very centralised system that revolved around the central police station. All major units and squads were located here, including the district head, the investigation units, the central switchboard, and most of the troops. Relations between the different echelons were direct. Even though patrols had considerable autonomy (Ericson 1982), no-one was the 'owner' of any one territory and many different patrollers intervened in the same place (Faivre 1993).

This was not the case in the *gendarmerie* system. Each BT team had a monopoly regarding all kind of interventions: emergency calls, road safety control, prevention operations, meetings with local authorities, criminal investigations up to a certain level, and all means of obtaining information. The same officers were asked to achieve all these missions. This versatility led them to interpret how to implement the orders they received from above and to choose their own priorities. They practically set their own policies by themselves, in cooperation with their local partners. A *gendarme* coming from outside spent his first weeks following an older colleague to learn the specific habits of his new territory. In this way, *gendarmes* participated in preserving the distinctive local characteristics of each part of France, within a centralised system (Emsley 1999).

The necessity of policing a huge rural territory (550,000 km^2) led to a system whose actual organisation went against the logic of centralisation which is the usual feature of French policy. In this system, senior *gendarmerie* officials mainly required that the BT provided reports and statistics, avoided large problems, responded to official demands, and informed the government of social unrest or important criminal cases. If they succeeded, the BT remained relatively independent.

Living Where You Work

To maintain control throughout the French territory, the authorities never questioned one principle: *gendarmes* were obliged to live in barracks located in the place where they work. Because they belong to the armed forces, *gendarmes* are not only considered as police officers, they are also supposed to protect the territory against foreign attacks. However far-fetched and inefficient that claim actually is, the outcome is that they could quit their workplace without official authorisation, on pain of punishment for treason. The six or eight *gendarmes* belonging to a BT therefore lived with their family in the barracks near their office (Clément 2003). Until the late 1980s, *gendarmes* led a traditional family life; *gendarmes* were married[2] men[3], their wives were housewives and lived in the barracks[4] (Bergère 2004). Although this made everyday life in the barracks rather difficult, it did oblige the six or seven families of *gendarmes* to establish contacts with other citizens. *Gendarmes* were therefore enmeshed in the local environment, not only because of their function, but also through their family.

Of course, as an important part of security policy, their numerous tasks led them to develop various types of relationships with the population. They knew both victims and offenders. But we must keep in mind that the *gendarme* was also a resident, and was therefore involved in the local community. He, or his wife, used the same services as other people; they had the same leisure activities; their children went to the same schools. The *gendarmes*, far from being schizophrenic, knew no border between their working time and their private life: they shared the same information and the same concerns as everyone else. In spite of, or because of, their military status and assignment to barracks, they automatically participated in the local community. Given this implication, it is not at all surprising that *gendarmes* gave preference to local priorities over national ones.

But this also had negative implications. In some cases, *gendarmes* clearly went along with rumours and charged people who were considered marginal by the majority of the population (such as travelling people), without sufficient proof of their guilt. This kind of excess was not unusual, including for serious crimes, as shown by several homicide cases[5] (Matelly 2000). One could also contend that this involvement makes *gendarmes* fight people who try to overthrow the established order or who question their fairness.

But conversely, non-implication in local community life is not synonymous with a lack of prejudice. Indeed, *gendarmes* with a strong residential connection were more inclined to make more holistic appraisals of people and communities:

> We're social workers as well. Our job isn't just to punish people. Our job is to
> help them. (*Gendarme* – Rural Territorial Squad – 1997)

----

2   Cohabitation was not authorised in *gendarmerie* squads until the mid-1980s.

3   Women were not allowed to enter the profession until 1984, and the number of women *gendarmes* was limited to 7% of the personnel for several years. At present, women represent 15% of all *gendarmes* (Source http://infos.emploipublic.fr/).

4   Marriages had to be authorised by the hierarchy until 1978. Wives were not allowed to work.

5   Such as the Roman Affair or the Pleine-Fougère murder, two cases played up by the media. In the first case, people accused a man on the fringe of the community of having raped and killed a seven-year-old girl, in 1988. He was acquitted in 1992 when the judges demolished an investigation conducted by *gendarmes* who were 'visibly influenced by their prejudices'. The second case was similar. In 1996, a young English woman was raped and killed. Here too, a homeless man was accused of the crime, to which he confessed after 45 days in police custody. Genetic fingerprinting later proved him innocent.

In their view, they had to understand people before charging them, especially those suspected of petty offences. And, in such cases, they often preferred to lecture offenders rather than sending them to court.

> You have to manage to solve problems without hurting people's feelings or pressuring them. Sometimes you have to know when to look the other way, not to dumbly enforce the rules, to be indulgent (that is, remove sanctions). Acting right earns us a degree of respect, and some help, because without the population we in the *gendarmerie* can't do very much! (*Gendarme* – Rural Territorial Squad – 1997)

That was their strategy with youngsters in particular. As a result, many of these offenders felt indebted to them for not having been prosecuted. This was also true of drivers who committed traffic offences: many were simply reprimanded rather than punished if they belonged to the local community (Zaubermann 1998a, 1998b).

Of course, the *gendarmes'* involvement in local social life may have led them to overstep the limits set by the law. Their ability to solve problems and their tolerance may have appeared as weakness. The boundary between cooperation and corruption was sometimes hard to define. Good relationships with the neighbourhood, especially when the citizens appreciated the *gendarmes*, often meant gifts, bottles of wine, facilities in obtaining some services or help in case of troubles.

From the *gendarmes'* viewpoint, accepting gifts from the population did not mean that they were corrupt. It was a 'normal' way of living amidst people and a requisite for integration in the local community (Luc 2002). For them, doing a favours for a neighbour in return for information was a logical step in the fight against more 'serious crime'. For many *gendarmes*, their involvement in the community was essential for achieving their mission, not for earning more money. Moreover, those who worked in intimate communities were more concerned than their colleagues who worked in more anonymous cities. Until the end of the 1990s, they were required to answer all phone calls or solicitations pertaining to their territory and it was out of the question for them to refuse to answer. Even during their leisure time, if something happened, they had to take action; otherwise they would have been totally discredited in the eyes of all the citizens.

Yet, chief officials could not accept any loss of authority that may have arisen from the close relationships between local people and 'their' *gendarmes*. Consequently, they imposed greater mobility on their staff in the 1990s, drawing upon old regulations that required *gendarmes* to change their place of work more often.

Lack of Resources

One would imagine that a centralised administration anxious to reinforce the chiefs' authority over rank-and-file *gendarmes* would have managed to find the needed resources to better control their troops. That was not true in the case of the *gendarmerie*.

Because the Ministry of Defence was primarily concerned with warfare rather than policing and the Ministry of the Interior were reluctant to finance staff from another ministry, the *gendarmerie* was often the poor relation when it came to finances. Furthermore, given that there were 3,500 BT across France, most funding went on salaries rather than equipment. *Gendarmes* were known to have old-fashioned cars, poorly furnished offices, and problems with their computers. Although the administration provided accommodation for *gendarmes* and their family it was unable to afford a modicum of upkeep.

The tradition has always been that it is up to *gendarmes* themselves to find the resources they need to work efficiently (until the 19th century, they had to buy their own horse to make patrols) and so it was in the interest of local gendarmes to develop close links with local mayors and other partners. They were obliged to ask the local authorities for funds for repairing their homes. The problem was similar for official cars, computers, or even to get paper to make reports. The central administration failed to supply enough resources to prevent their subordinates from depending on those local authorities.

Face with this lack of resources, the *gendarmes* never gave up. They asked their partners for what they needed to work and live properly. For a while, local authorities paid for the maintenance of *gendarmes*' houses or flats. Town clerks and secretaries of private firms knew *gendarmes* well, since they were always asking for paper, pens or printers. Close relations with garage owners enabled them to repair their cars. Of course, the constraint created by this obligation to ask for resources put *gendarmes* in a position of dependence. It is not easy to punish someone and then ask him for goods, or to refuse a service to a mayor and then file a request for repair work on the roof of the *gendarmerie* building:

> We ask the local print shops to give us reams of paper. We go crying to the local garage to get spare parts. We're beggars. That puts us at a disadvantage: people think the *gendarmerie* is under their heel, and they make it clear that if they do you a favour, you have to return it. (*Gendarme* – Rural Territorial Squad – 1997)

If the *gendarmes* suffered from this situation, they also found the 'poor guy' role somewhat advantageous. As one of the *gendarmes* interviewed in our fieldwork confessed, '*it is sometimes useful to have something to ask for when you want to get in contact with somebody*'. Many *gendarmes* explained how they used their request in order to approach people, and to show that they were unassuming, not dangerous, and anxious to create relations based on mutual aid. They avoided arousing distrust, and incited the people to give them information. For example, when asking a garage owner to repair their cars, the *gendarmes* could ask about damaged cars which might have been involved in a hit-and-run accident. When they begged for paper in a factory, they could engage in a discussion about the social climate and, if necessary, work out a security device. This forced modesty was turned into an advantage: it helped the *gendarmes* build a network in different circles.

The *gendarmes*' strategy in the local environment was therefore a mix of making demands and doing favours. Because they had a large territory to control with little forces, they needed a dense network of people ready to give them all kinds of information. In every village, in every social circle, there was always at least one person, be it a peasant, a shopkeeper or an old woman, with a direct link with a *gendarme*. Although neighbourhood watch has officially never been recognised in France, the *gendarmes* have nevertheless invented it. And when *gendarmes* were looking for elements in a judicial case, those 'correspondents' gave them everything they wanted.

In order to maintain these links with all the people in their hidden but real network, *gendarmes* used two main strategies. First, they tolerated some rule-breaking by those people. Second, they themselves accepted a social worker role: they made frequent visits to the elderly; they tried to find solutions for people who had problems with some administration. In that sense they were spontaneously applying the problem-solving policing strategies that have received theoretical formulation elsewhere (Goldstein 1990). How much of this was personal interest, how much part of the job? It is hard to define, because the *gendarmes* were totally involved in local society, and proud of being of help to citizens.

From 'Old-fashioned' *Gendarmerie* to Modern Bureaucracy

Although the *gendarmerie* was all-pervasive, efficient and probably instrumental in protecting the countryside, it has been criticised as old-fashioned, expensive and unable to face up to new societal challenges. New productivity measures were introduced that questioned the efficiency of the *gendarmes*. This not only affected *gendarmes* but also their families who no longer accepted the burden imposed on them.

In 1989 the first internal 'revolution' broke out: *gendarmes* sent anonymous letters to the Head Office of the *gendarmerie*, complaining about the burden of their work, the lack of understanding expressed by their superiors, and their living conditions. Using this as an excuse, Head Office imposed a new system, patterned after the *Police Nationale* model, albeit the *Police Nationale* was concurrently attempting to improve community policing by applying a model very similar to the old *gendarmerie* one! A second rebellion, in November 2001, marked another turning point: the *gendarmes*, who as military men do not have the right to demonstrate, took to the streets to protest against their living conditions and their superiors (Cleach 2007).

The Causes of the Crisis

The equilibrium between central administration and local involvement was broken at the end of the 1980s as the hierarchy tried to reinforce its control through more rules and a more authoritarian attitude.

The *gendarmes* themselves were also part of the evolution because of the changes they felt in their own families, they no longer wanted to be so involved in their community. Their spouses (who had begun to work years ago) and their children were no longer ready to accept being disturbed day and night (by sometimes trivial phone-calls), being constantly watched and scrutinised by the other inhabitants and living in barracks where they were under the surveillance of the head of the BT and his wife (Clement 2003).[6] In addition, the families of these *gendarmes* found it difficult to live so far from an urban centre with modern living conditions and easy access to health, cultural and educational facilities. In 2001, our interviews with the *gendarmes* in the barracks of the most depopulated areas clearly showed how completely they rejected their assignment to the rural milieu.

> Just look! There's nothing here! You have to do a half-hour drive even to go shopping. You can imagine my wife's face when I told her we were going to live here for two or three years. (*Gendarme* – Rural Territorial Squad – 2001)

In short, *gendarmes* and their families wanted to set boundaries between their private and working lives. They started looking enviously at the police officers from the *Police Nationale*, who, as civilians, only worked a 40-hour week and who had unions to defend them.

This movement was encouraged by social change in rural localities. *Gendarmes* work in two kinds of rural areas: those suffering from a population drain and those in the process of urbanisation. In areas losing their population, the *gendarmerie* was the last remaining public service, funded by the government for the sole purpose of controlling rural places. As such, they not only policed crime but carried out broader social functions, including the production of local statistics, monitoring social movements and to informing the authorities how a heat spell affected the elderly.

> We're the administration's maid of all work in the countryside. We're asked to do everything, from monitoring the level of the rivers to checking the supply in

---

6   Between the 1989 and 2001 'movements', several campaigns were launched featuring 'letters from *gendarmes*' wives' sent to the Ministry of Defence, to show their discontent.

gas stations when the teamsters are blocking the refineries, and to making sure the elderly don't die. (*Gendarme* – Rural Territorial Squad – 2001)

At the same time, the rural population has tended to make all sorts of new demands on *gendarmes*, especially in areas with elderly residents (Desplanques 2005) where the *gendarmerie* are seen to compensate the absence or shortcomings of the other administrative services:

It's not unusual for us to take a special delivery letter to an elderly person who isn't able to go out. That's common for us. (*Gendarme* – Rural Territorial Squad – 2001)

In more accessible rural areas, the inhabitants are no longer agricultural workers[7] but people who work in the cities. Thus, their social, working and commuting structures vary considerably from the traditional French rural communities which tended to function on the basis of automatic solidarity (Durkheim 1893). In the towns closest to the cities, low-rent public housing has been built and many residents are poor. The location of barracks in these localities can be a source of tension between residents and *gendarmes*. Thus, some *gendarmes* were reluctant to send their children to the same schools, fearing they would get a low-level education in classes with few middle-class pupils.

Increasingly, fewer *gendarmes* are from a rural background themselves (Mucchielli 2007) and so are reluctant to live in a rural environment they find boring and poorly serviced. Even experienced officers are deserting the rural world for places close to a city. Since they are recruited on a national basis, holes in rural provision are filled by the latest recruits. As a result, all new recruits are sent to the least-densely inhabited areas but, two or three years later, they move to other BTs that they perceive to offer a better quality of life for their families. More and more *gendarmes* now prefer to live near (but not in) cities because their partners are working there and they offer the best schools and facilities for their children. This desire reinforces the idea of a clear separation between work and leisure. Some *gendarmes* have decided to install their family in a house far from the barracks and to stay in the latter only when on duty. Overall, *gendarmes* who transfer are less likely to be involved in local life

The current trend toward modernisation of the corps and the persistent determination of the hierarchy to reinforce its power over subordinates will help the *gendarmes* achieve their objective. The first step was the creation of a central switchboard, one per county, centralising all phone calls and demands made on the BT intercoms. At first, the *gendarmes* were very happy because they were not disturbed as often but it meant a loss in autonomy. The centre was located near

---

7   Working farmers represented 5% of the French labour force in 1990, as opposed to 27% in 1954. In 1990, farmers represented 18% of the rural French population, as opposed to over 60% 100 years earlier (source: INSEE).

the highest county-level officer (the colonel) who then controlled requests and ordered subordinates what to do.

The second step was to replace the BT patrols by car patrols coming from, and in constant contact with, regional headquarters. Again, the *gendarmes* were at first satisfied from the exemption of this work, but after a while they came to understand that they had lost their monopoly over relations with the community. More generally, BTs are now accountable to their chief officer for all their actions and, consequently, more reliance has been placed on performance data. The main concern is no longer keeping the peace in the community served by the BT but achieving performance targets. It is now out of the question for a patrol to spend time solving social problems or talking with the population: contacts must respond to logics of efficiency, measurable through statistics (Matelly and Mouhanna 2007).

Consequences and Conclusions

The statistics-based management of the *gendarmerie* is grounded in the New Public Management theory. The idea is to optimise the adjustment of resources to problems encountered locally. To do so, the plan, theoretically, is to eliminate some positions when the work load is light. The problem is that the indicators chosen to measure police activity and efficiency are based exclusively on the number of complaints registered and the elucidation rate. The first result of this policy is a withdrawal from all those activities that are 'not profitable': there is no accountancy of the *gendarmes'* participation in community life or of their unofficial problem-solving because no indicator measures those activities. The second effect is that *gendarmes* play the numbers game themselves. For instance, they try to dissuade people from filing suits that will impact negatively on their statistics and they conduct more road checks to 'improve their figures', which annoys the population. The third consequence is organisational: each *gendarme* attempts to meet the objectives assigned by the hierarchy and no longer listens to people's demands (Matelly and Mouhanna 2007).

What this means is that figures are of little use in determining whether the *gendarmes* in any particular squad work more than some other one, as illustrated by one of many interviews with the head of a BT:

> You can make statistics say whatever you like. You change the charges in a case so as to lower the figure for offending. For a squad commander, that's both good and bad. The advantage in lowering offending is that the media and your hierarchy approve. The disadvantage is that if offending declines, your squad is cut back. The chief has to be intelligent enough to reconcile the two contradictory goals. If offending increases nationally, local figures have to reflect the national level. But you mustn't do that stupidly, you have to take specific parameters into consideration, such as the opening of a high school, which brings in more

rackets and drugs, or a new suburb, which increases the number of thefts and attacks. The squad commander has to anticipate all those events, and infer a future rise in offending in his district. He needs more men to respond to future needs, so he is prepared to have bad statistics, by showing an artificial rise in offending. (A Commander of a Territorial Brigade)

One thing is clear, however: that is, the *gendarmerie* hierarchy, and above it the Ministry of the Interior (which has replaced the Ministry of Defence in administering the *gendarmie*), prefers management using national indicators to direct management by the local hierarchy. Rather than relying on the judgment of the latter to assess the efficiency of a BT, through exchanges with local elected officials in particular, there is a preference for reliance on figures that ignore the dimension of relations with the public. As for human resources, transfers are encouraged: a long stay in the same BT and integration in the local community are no longer viewed as qualities. Here again, *gendarmes* are kept away from the public.

The pressure placed on French public agencies by the New Public Management policy has increased the gap between the *gendarmerie* and rural citizens. The Ministers in charge of safety have planned a vast reorganisation of the BTs. One third are, or will soon be, merged with a larger one. Others, near cities where the *Police Nationale* is responsible for security, are being included in the police territory and closed.

Gendarmes who want to maintain their tradition find it increasingly difficult to continue as before. The *gendarmerie* has entered the same vicious circle that has characterised the French *Police Nationale* for decades: police officers work mainly for the priorities set by their superiors and spend more time controlling the population than serving it. They make short shrift of the population's needs (Monjardet 1996), and have therefore lost its confidence: they are viewed as unnecessary, or even as enemies. They have little contact with citizens, especially youngsters, which reinforces mutual fear. The use of technology, rather than personal contact, as an answer to every problem has produced a gap between the French police and the population. It has become a bureaucracy, in which the needs of the public are ignored and priorities depend on the government's orientation and on corporatist demands. As of January 1, 2009[8] the *gendarmerie* was placed under the authority of the Ministry of the Interior in spite of opposition from both left and right-wing local elected officials.[9] This is the last step in the *gendarmerie's* evolution from community policing to a national force that is based on control rather than public service (Monjardet 1996). The repeated riots in the French

---

8   The bill was voted on July 7, 2009, six months after the actual transfer of the *gendarmerie* to the Ministry of the Interior.

9   See, for instance: Fusion police-gendarmerie: la revanche de Fouché ? Notes – Jean-Jacques Urvoas – 26 juin 2009, http://www.tnova.fr/index.php?option=com_contentandview=articleandid=846.

'*banlieues*' (suburbs) is one outcome of this policy. Local authorities, mayors, and county councils attempt to compensate for this desertion by creating new public services to regulate community life.

The situation in present-day rural French society seems to be as complicated as in the cities, but with less resources. Indeed, the *gendarmerie*'s withdrawal reflects a far-reaching movement encompassing much of what the French call 'public services', provided by the state or by publicly-owned companies, including post offices, hospitals, and roads. Because they are not considered profitable, all these services are being reorganised and often closed. Local authorities do not have the resources needed to replace them and people from rural areas feel they have been abandoned. There are many indications that fear is increasing. The *gendarmes*, like other civil servants working in these areas, took part in building a nation, that included these outlying towns and villages, by providing security and other services (Emsley 1999). Today this is no longer the case. It remains to be seen whether a vast part of the countryside will become a lawless territory or, instead, if local communities will create new forms of policing in which the state plays a very small part.

Chapter 6
Plural Policing in Rural Britain

Rob I. Mawby

Introduction

The last 20 years has seen an expansion in the number of private and state agencies involved in policing (Bayley and Shearing 2001). Although much attention has been given to the rise of private policing of urban space (de Waard 1999; Jones and Newburn 1995), their influence in rural areas is less significant. Instead, the use of volunteers and police ancillaries has been more significant in the development of the rural policing mix.

While a plethora of authors have focused on the expansion of multilateralised policing, none address the spatial components of this shift. Indeed, three assumptions appear implicit to much of the discourse: firstly, that expansion in the range of providers is universal, both between and within societies; secondly, and following from this, that expansion is similar in both rural and urban areas, with a resulting similarity to the policing mix; thirdly, that the different forms of multilateralism have developed in similar ways. In challenging this view, this chapter demonstrates that the policing mix is very different in rural and urban areas. Given the paucity of international data in this respect, the British Isles is used as an example, utilising three case studies: national statistics for England and Wales on police force areas' personnel; data from Jersey, the largest of the Channel Islands, on the distinctive structure for police volunteers on the island; and findings from the Cornwall Crime Survey (CCS) and Cornwall Business Survey (CBS), on public and business people's perceptions of policing alternatives in this rural county. Before that, however, since much of the discussion regarding the policing mix has focused on private policing, volunteers and policing ancillaries are considered in more detail.

Volunteers

Local communities still seem more likely to rely on self-help than turn to the private sector and the most common form of crime prevention is to ask neighbours to keep an eye on one's property (Zvekic 1998). Yet spontaneous or semi-spontaneous groups may be established in reaction to an ongoing concern and may achieve some degree of permanence, structure and pro-active operation.

Thus, local communities may engage in vigilante action (Johnson 1996) that comprises some degree of organised action by private citizens, using force or the threat of force to impose control and restore security. Although more a part of the US tradition, it is also evident in Europe and attempts by hunt supporters to 'protect' their sport from protesters provides a rural example. Such actions rarely have police support (see Woods, this volume).

In contrast, local people may be recruited by the public police on an unpaid voluntary basis to provide policing services. Neighbourhood watch is a case in point. Despite limited evidence that they have any impact on crime rates (Bennett 1990; Laycock and Tilley 1995) or fear of crime (Rosenbaum 1988; Sims 2001, 2003), neighbourhood watch was enthusiastically endorsed by the British public (Yarwood and Edwards 1995). Although 27% of households in England and Wales described themselves as members of a scheme in 2000 (Sims 2001), there has been little evidence of any sustained growth in neighbourhood watch participation since then and questions on involvement have been quietly dropped from recent British Crime Surveys.

It might be expected that the gemeinschaft qualities of many rural areas, with greater potential for social capital, would mean that neighbourhood watch would be more evident in such areas. This is not, however, the case. BCS data suggest that while schemes are least common in inner city high crime areas, they are most likely to thrive in suburban middle class areas (Sims 2001). A survey by the author in Devon and Cornwall showed that schemes and participation were greatest in the most urbanised parts of the force area and least common in rural Cornwall. This is perhaps not surprising as, in addition to high social capital, there is a need for local residents to consider crime a particular threat if neighbourhood watch is going to be sustained. High crime areas may be weak in the former and strong in the latter respect, and low crime rural areas strong in the former and weak in the latter respect. Middle-class suburbs with low crime rates but in close proximity to high crime areas may be the most fertile ground for schemes.

A further form of public participation is the Special Constabulary (Gill and Mawby 1990a). This was introduced prior to the establishment of a public police and was used as a back-up to the military in periods of labour unrest, often in rural industrial areas of northern England. As such it was predominantly manned by politically safe, upper- and middle-class men. It continued after the establishment of the regular police, as a reserve deployed in times of public disorder. In the immediate post-war period, the mandate of the Special Constabulary in England and Wales and Scotland was radically altered and the Special Constabulary came to be seen – and used – as a form of community policing, a bridge between the regular police and public. Today it is part of the public police comprised of trained volunteers who commit at least four hours each week to police duties.

However, while the post-war target was to recruit at least one special for every two regular officers, the special/regular ratio never reached this. The numbers of specials steadily decreased from the early 1920s where they peaked at around 128,000 officers. Despite initiatives aimed at boosting recruitment (Gill and Mawby 1990a; Mirrlees-Black and Byron 1994), including attempts to rebrand them (Southgate, Bucke and Byron 1995), this downward spiral continued until

2005, since when there has been an annual rise. Interestingly, specials are more representative of the general population than are regular officers, including more women and ethnic minority officers. One notable geographical difference is in recruitment, with rural constabularies traditionally recruiting proportionally more Special Constables (Fyfe 1995; Gill and Mawby 1990a).

Policing Ancillaries

An increasingly common development in a number of countries has been the use of wardens employed by non-police state agencies to provide a policing function (Stockdale 2002). The concept of paid specialist non-police patrols on public transport and in commercial areas was established in the Netherlands in the 1980s (Hauber *et al.* 1996; van Steden and Huberts 2006), predominantly in metropolitan areas.

Two recent and similar developments in the UK are Police Community Support Officers (PCSOs) and Neighbourhood Wardens. Both are predicated by a concern to cut expenditure by deploying less well-trained and lower-paid staff to carry out less-skilled tasks that in the past were the responsibility of police officers. PCSOs, introduced following the Police Reform Act 2002, are uniformed civilians employed by police authorities to deal with lower level crime and anti-social behaviour, engage with local communities, and offer effective crime prevention advice. They spend a considerably greater proportion of their time on patrol than do regular officers (Cooper *et al.* 2006; Crawford *et al.* 2005). They do not have the power of arrest, but in some police authorities, may be designated with the power to detain suspects until a regular officer arrives (Singer 2004). The use of CPSOs seems to have focused, at least implicitly, on urban areas.

'Wardens' are employed within the public sector but outside the police authority structure (Jacobson and Saville 1999; National Strategy for Neighbourhood Renewal 2000). They provide a uniformed, semi-official presence and aim to improve the quality of life in depressed neighbourhoods by reducing crime and anti-social behaviour. Although wardens have no special powers of arrest, they provide a local and accessible presence, a focal point for the gathering of information about local concerns, to which they may respond directly or by referring matters on to other agencies, including the police (National Strategy for Neighbourhood Renewal 2000; Office of the Deputy Prime Minister 2004).

Evaluations suggest that wardens are effective, particularly in areas where the crime rate is high, anti-social behaviour is rife, and there are poor relationships between local people and the conventional police (Crawford *et al.* 2005). This implies that they might be better utilised in more urban areas, but some rural boroughs have also employed them.

Plural Policing Forms in Rural Areas of the British Isles: Three Case Studies

The Rural Policing Mix

Data for England and Wales (Bullock 2008) reveal the extent to which regular police officers, although still a majority of police staff, are nearly matched by other employees of the 43 Constabularies. Combining regular officers, civilians, PCSOs and Special Constables, regular officers accounted for 57% of the total, with civilians accounting for 31%. PCSOs, and Special Constables, while less common, each accounted for about 6% of the total. Put another way, by 2008 there were 11.2 PCSOs and 10.4 Special Constables per 100 regular officers. The balance has also changed in recent years. After fluctuating in the late 1990s the number of regular officers increased from 2000 to 2006 with increased central government funding, before falling back in 2007 and 2008. However numbers in 2008 were still 110.5% the 1996 figures. In contrast, the number of Special Constables declined from 1996 to 2004, continuing the trend of earlier years (Gill and Mawby 1990a), but then rose as central government financed a number of recruitment initiatives. Nevertheless, the number of Special Constables in 2008 was only 73.6% that of 1996. The more recent rise in the number of Special Constables broadly coincided with the introduction of PCSOs in 2003. In that year there were 11,037 Special Constables and 1,176 PCSOs. By 2008, the number of Special Constables had risen to 14,547, an increase of 31.8%. These were, however, outnumbered by PCSOs, whose numbers had been increased more than tenfold to 15,683. Thus, rather than their being a cross-the-board increase in alternative forms of plural policing, we can see quite distinct patterns, with PCSOs apparently replacing Special Constables.

But this national picture is somewhat misleading. The extent to which conventional police officers are supported by other forms of public policing varies markedly between force areas. The number of regular officers per 100,000 population tends to be greatest in those Constabularies catering for metropolitan areas, with the highest serious crime rates. The formula used by the Home Office to allocate resources, based as it is upon crime data and other demographic data indicating stress, effectively leads to greater funding for the metropolitan forces, with consequently more regular officers being deployed. Thus in 2008 the five forces with the greatest number of regular officers per 100,000 population were the London Metropolitan Police (430), followed someway behind by Merseyside (331), the West Midlands (324), Greater Manchester (315) and Cleveland (303). At the other extreme, the constabularies with the least regular officers were Lincolnshire (175), Surrey (179), Cambridgeshire (183), Suffolk (188) and Wiltshire (190).

When PCSOs were introduced, there was an implicit emphasis upon their appropriateness in urban areas, with the less numerous Neighbourhood Wardens, employed by local government, seen as possibly appropriate for rural as well as urban, policing support. In 2008, PCSOs were more common in more metropolitan, 'big budget' constabularies with high crime rates: the Metropolitan Police employed by far the most PCSOs per 100,000 population (58), with Cleveland (30), Greater Manchester (30), Merseyside (30) and the West Midlands (29) also prominent, albeit these were overtaken by Humberside (35), West Yorkshire (35) and Norfolk (33), the latter being the only largely rural constabulary ranking highly. At the other extreme, PCSOs were least common in Dyfed Powys (15), Derbyshire (17), Hampshire (18), Northumbria (18) and Surrey (19). There is also a strong and statistically significant positive correlation between rates of regular officers and PCSOs (r=0.675, p<0.01).

The fact that Special Constables are more commonly deployed in more rural constabularies has long been recognised (Gill and Mawby 1990a; Fyfe 1995) and is at least partly a factor of a tradition of voluntarism in such areas (Gill and Mawby 1990b). Not surprisingly, then, the constabularies with the most Special Constables are relatively rural: Suffolk (51), Derbyshire (45), Dorset (42) and Warwickshire (40), with Humberside (38), ranked fifth the most notable exception. The group of constabularies with least Special Constables is more mixed though, including: Northumbria (26), Thames Valley (110), Sussex (13), Greater Manchester (14) and West Yorkshire (19). Finally, there is no relationship between the rate of Special Constables and either regular officers or PCSOs.

The relatively close relationship between regular officers and PCSOs means that variation between forces in the number of PCSOs per 100 regular officers is far less (5.54–17.37) than for Special Constables (3.09–27.31). If these figures are taken as crude indications of the extent to which forces use PCSOs or Special Constables rather than regular officers, a distinctive pattern emerges for the latter. The number of Special Constables per 100 regular officers is greatest in Suffolk (27), Derbyshire (21), Warwickshire (20), Dorset (19) and Norfolk (17), all relatively rural. In contrast, the PCSO ratio is high in some more rural forces, such as Norfolk (17), Cambridgeshire (14), and Lincolnshire (14) but also for Humberside (14) and the London Metropolitan Police (14). There is no overall relationship between the two ratios. That said, some extremes can be identified. For example, Derbyshire, ranks second on the Special Constables ratio and has a low ratio of PCSOs (8), whereas the Metropolitan Police, with a high ratio of PCSOs, ranks low for Special Constables (8). On the other hand, rural Norfolk is ranked first for PCSOs and fifth for Special Constables. Combining the two, both Norfolk (35) and Suffolk (38) have more than one PCSO or Special Constable per three regular officers. The 'outliers' appear to be Humberside, a relatively urbanised constabulary with a high proportion of Special Constables, Norfolk, a more rural constabulary with a high proportion of PCSOs, and Thames Valley and Sussex, with less Special Constables than one might have expected.

Of course, these figures take no account of the number of personnel engaged 'on the ground' in policing at any one time. For example, PCSOs are expected to spend relatively more of their time 'on the streets' than regular officers, whereas Special Constables, working a minimum of four hours per week, have less working hours available for patrolling. Nevertheless the figures do indicate a pattern, with large metropolitan forces heavily reliant on regular officers, increasingly supported in some cases by PCSOs, and more rural forces, with proportionally less regular officers, making more use of Special Constables and, in some cases, PCSOs. As a result, policing in the countryside involves a somewhat different mix of personnel than urban policing.

The Policing Mix on Jersey

Covering 45 square miles, and lying only 15 miles from the French coast, Jersey is the largest of the Channel Islands. It has a population of some 85,000 people and its capital, St Helier, houses about 30,000 of these. The importance of police volunteers in rural communities is reflected in policing structures on the island.

Although Jersey came under the English Crown following the Norman conquest, its government has remained distinctive. Its 12 parishes form the basis for its social and political development and, subsequently, its criminal justice system. One notable feature of this is its volunteer police, known as the Honorary Police System, that is parish-based and largely autonomous of the public (States) police. This system preceded the establishment of a public police force in the late 19th century but, unlike on mainland Britain, maintained its independence following the introduction and extension of the public police, in some respects holding more power than the latter, being closely integrated into the parish political structure (Mawby 1999).

Jersey's professional police service, the States Police, is comprised of approximately 250 officers, headed by a Chief Officer. There are 2.75 police officers per 1,000 population, well above the average for England and Wales, (States of Jersey Police 2005), and in marked contrast to rural forces in England and Wales. The main police station is in the capital, St. Helier, although there is also a station at St. Brelade's that is shared with the Honorary Police. Most officers, though, operate out of St. Helier.

There are currently about 300 members of the Honorary Police, comprising three ranks: Centeniers, Vingteniers and Constables' Officers. The Connetables, the heads of government in each parish, were until recently formal heads of the parish Honorary Police, although in practice most Connetables had little to do with police management, preferring to delegate to Centeniers (States of Jersey 1996). Following a review of the system in 1996, the Connetables were stripped of their responsibility for the Honorary Police and in each parish a Centenier, with the formal title of Chef du Police, is now responsible. There are about 48 Centeniers who are on duty for a week at a time and on call for 24 hours. Only the Centenier is empowered to charge or bail suspects. Centeniers also play a key role in the prosecution system.

Below the Centeniers come the Vingteniers, of whom there are about 60, and the Constables' Officers, numbering some 174. Although Clothier (States of Jersey 1996: 14–15) felt that Vingteniers had little authority, they might perhaps be equated with the duty sergeant, with responsibility for a shift comprising Constables' Officers and one or two Vingteniers.

Two parishes can be cited to illustrate the operation of the system. At the time of Gill and Mawby's (1990a: 81–2) research, in St. Peters, a medium-sized parish in which the airport is located, the Honorary Police were based in the Parish Hall, using a marked police car with the insignia of the parish to distinguish it from the cars used by the States Police. The parish had four Centeniers and five Vingteniers, each responsible for a 'watch' of four Constables' Officers. Each 'watch' carried out duties for a week on a rotation basis.

The St. Helier Honorary Police, in contrast, were based in the most urban parish on the island. There were 10 Centeniers, 11 Vingteniers and 20 Constables' Officers. The Centeniers were on duty for a week at a time. The Vingteniers and Constables' Officers worked in four teams of seven to eight, also for a week at a time. They concentrated on town patrols in the evenings at weekends, and additionally patrolled in the week when necessary. During 1997 they carried out 9,615 hours of duty and made 84 arrests.

Clearly these examples illustrate the extensive involvement of the voluntary police, especially in comparison with their English counterparts, the Special Constabulary (Gill and Mawby 1990a). Recent changes have curtailed both the autonomy and power of the Honorary Police (Rutherford 2002b; States of Jersey 1996), but although the Honorary Police has faced recruitment difficulties, it retains a visible local presence and is a significant part of island policing.

The continuing importance of the Honorary Police can be illustrated in six ways. First, in terms of their numeric significance, volunteers outnumber paid police. Secondly, in terms of structure, the Honorary Police, unlike the Special Constabulary in England and Wales, is a quite separate entity and is not controlled by the States Police. Thirdly, in terms of the nature of the work, the police volunteers have a primary responsibility for routine traffic checks, covering drink driving, vehicle roadworthiness etc., as well as handling firearms licenses. Fourthly, in terms of the law, the States police are obliged to inform the Centeniers when they are called to incidents within the parish; Centeniers act as magistrates for minor crime and prosecutors in the Magistrates Court; and the legal powers of Centeniers (regarding charge, bail, and search without warrant) exceed those of the regular police. Fifthly, in terms of accountability, the regular force is accountable to the Home Affairs (formerly Defence) Committee on which the Honorary Police, as part of the political structure of the island, still exert an influence. Finally, in terms of community-orientation, Honorary Police are based in the Parish Halls, using marked cars paid for by local taxation, whilst the involvement of the States Police in communities outside St. Helier has historically been restricted, making it more of a reactive force.

Jersey thus provides an excellent contemporary example of a rural area in which community involvement in policing has been maintained up to the present. However, there is little evidence that this community presence is growing. Indeed, on the contrary parishes are finding it increasingly difficult to recruit to the

Honorary Police. Moreover, although the system may be seen by some as utopian, it also evidences many of the problems associated with too great a dependency on the local community. Recent examples include cases where the Honorary Police have turned a blind eye to domestic violence within 'their' parish, problems of poor communication where local complaints have not been passed on to the States Police, and claims that local pressures on the Honorary Police make it difficult for them to act impartially.

Policing Alternatives to Tackling Crime and Disorder: The Case of Cornwall

This section is based on surveys conducted in Cornwall in 2004 as parts of the local crime and disorder audits (Mawby 2007, 2009). Cornwall is known to most people as a tourist base, but is also characterised as a rural county with high levels of poverty and deprivation. Only nine towns in Cornwall have populations in excess of 10,000 and its population density is 0.1381248 people per 1,000 hectares.

While in general people living in rural areas tend to be more positive about the police than urban dwellers (Mawby 2004, 2009), a significant proportion of both residents and business people in Cornwall expressed criticism of the public police. The imposition of a centralised call system, a reliance on patrol cars and the gradual closure of local substations, has highlighted the problem of access to the police (see Mawby, Chapter 2 this volume). In the light of such concerns, it might be thought that people would welcome plural policing options that gave them a greater access to alternative policing agents of one sort or another. What, then, did local people think of alternative policing options currently available?

Community/neighbourhood wardens and PCSOs provide alternative publicly funded options; in each case they have been adopted in Cornwall and appeared to be overwhelmingly endorsed by the public. No less than 80% of the public and 75% of business people thought that appointing community wardens or Police Community Support Officers in Cornwall to work with the police and other agencies to help 'with local problems, including crime and disorder' was a good idea.

Neighbourhood watch received an even more positive endorsement, with 90% considering it a good idea. However, only 29% said there was a neighbourhood watch scheme active in their area. Not surprisingly, even less said they had worked in the other main voluntary policing area, as a Special Constable. However, far more said that they had made a less formal contribution. For example, 55% said they had 'kept an eye on a friend's property when they are away'.

Business people were also asked about security precautions. Whilst few (6%) were members of a business watch scheme, 15% employed in-house security and 11% used an outside security firm. Overall, then, it appeared that whereas local people commonly asked neighbours to keep an eye on their property, and a significant minority of businesses deployed private security, other examples of plural policing were far less common.

Respondents were also questioned about what could be done locally to reduce car crime, burglary, violence, and disorder. In all, 23 alternative measures were included. Six of these involved police and policing. The first two options (more police on foot; more police in cars) related to increases in conventional public policing; the next two (council operated community patrols; more use of community wardens) to alternative public sector patrols; and the final two (more neighbourhood watch schemes; more private security) to greater involvement of the community and private sector.

Table 6.1 indicates that, whatever their criticisms of the public police, Cornwall residents felt that they constituted the most effective way of reducing crime and improving public safety. This was particularly the case for foot patrols, but applied to a lesser extent to vehicle patrols, the main exception here being burglary, where the advantages of rapid response (Coupe and Griffiths 1997) were clearly not appreciated by the public. Neighbourhood watch was, as in England and Wales as a whole (Sims 2001) seen as particularly effective in combating burglary, being outranked only by 'better home security'. Other public sector alternatives, especially community wardens, also received considerable public support. However, private security received less support. It ranked last of the six policing options in three cases, although it was considered marginally more useful in the case of burglary. The replies from business people were broadly similar.

A scale based on combined responses to the four crime and disorder problems cited was then constructed, such that an individual who felt that a particular measure would help reduce car crime, burglary, violence and disorder would score four for

Table 6.1   Percentage of local residents who felt the following would help reduce ... (rank out of 23 in brackets)

	Car Crime	Burglary	Violence	Disorder
More police on foot	63.9 (1)	61.9 (3)	69.1 (1)	75.9 (1)
More police in cars	46.8 (6)	36.0 (14)	38.6 (8)	45.0 (9)
Council operated community patrols	38.9 (10)	39.7 (12)	31.5 (11)	42.2 (12)
More use of community wardens	42.2 (8)	42.5 (8)	34.3 (10)	43.3 (11)
More neighbourhood watch schemes	39.0 (9)	62.8 (2)	21.1 (17)	24.0 (17)
More private security	23.3 (17)	39.3 (13)	11.5 (18)	12.4 (19)

Table 6.2   Mean scores for each of the six policing options (CCS)

More police on foot	2.71
More police in cars	1.66
More use of community wardens	1.62
Council operated community patrols	1.52
More neighbourhood watch schemes	1.47
More private security	0.86

this option. The mean scores here again demonstrated the relative popularity of foot patrols and reservations about private security (Table 6.2).

To summarise: 'multilateralised' policing may provide possibilities for additional policing presence in rural areas, and different ways of providing a visible policing presence on the streets indeed gained widespread approval in Cornwall. However, there was no evidence that the public was rejecting the public police and looking to alternative forms of policing instead. Rather, other options within the public sector, local community and private sector were seen as second (or third) best. What the public wanted was a return to their ideal of rural police services in the past, where police were more tightly enmeshed in their local communities (Cain 1973).

Discussion

Policing is no longer a 'near monopoly' of the public police. Indeed, in many countries it appears that only a small proportion of policing is carried out by trained, special police officers employed by central or local government. Discussion of multilateralised policing allows us to identify a number of alternative commissioners or providers of policing. However, it would be wrong to assume that these alternatives are emerging simultaneously, or that they are evenly distributed across or within countries. International data, for example, illustrates considerable variation between countries in the deployment of private security and the involvement of the public in the policing process.

A review of Home Office statistics for England and Wales illustrates the immense variations in the 'policing mix' between different forces. While the pattern is by no means clear-cut, it is evident that rural forces are able to call on a greater number of Special Constables than are the large metropolitan forces, more of which have become increasingly reliant upon PCSOs. Given that the number of regular officers is proportionally lower in rural areas, this means that the number of Special Constables relative to regular officers is greater there, a pattern that has remained consistent over the years, despite the expansion of multilateralism. The case study of Jersey illustrates the extreme. This rural island has relied on police volunteers since before the creation of a public police, and these volunteers continue to outnumber the States Police. However, as on the mainland, it is becoming increasingly difficult to maintain recruitment levels without considerable funding, demonstrating that it is misleading to assume growth of all forms of plural policing alternatives.

The assumption that social capital is greater in areas characterised by gemeinschaft qualities is thus a dangerous one, especially as the characteristics of residents of the countryside change. Moreover, as data from the Cornwall Crime Survey illustrate, local community support for plural policing alternatives is tenuous. For while people are strongly supportive of initiatives such as neighbourhood watch, they are far less enthusiastic about becoming actively involved themselves: hence the fact that the 1980s/1990s acceleration in neighbourhood watch has now stalled. Moreover, while local people supported the full range of plural policing options, they saw them as a second class alternative to the conventional public police. There is a real danger, then, that if rural areas become more reliant on, say, Special Constables or neighbourhood watch, residents will interpret this as evidence of a lack of commitment – by government and police agencies – and a lack of appreciation by them of the crime and disorder problems in the countryside.

Finally, though, it is worth emphasising that concerns over quality of service are as important *vis-à-vis* community or private sector involvement in policing as they are for the public section. Tellingly, surveys of police response to complainants on Jersey are restricted to the States Police and exclude consideration of the Honorary Police, and it is difficult to call to mind any public surveys that look at the services provided by either the private sector or local volunteers. Moreover, quality of service cannot be restricted to the views of complainants as consumers of policing services. It incorporates notions of the way suspects are treated and the effectiveness and efficiency of policing agencies in responding to crime and disorder incidents. The Jersey example suggests that in some respects the closeness of policing agents to their local communities is far from a panacea, and that justice may not be served where law-enforcers are too closely embedded in the communities they police.

Chapter 7

Governing Crime in Rural UK: Risk and Representation

Daniel Gilling

Introduction

The purpose of this chapter is to examine how rural crime is governed in the UK. In comparison to its urban cousin, the rural remains something of a 'criminological *terra incognita*' (Girling *et al.* 2000: 12). In the last few decades there have been significant changes in the local architecture for governing crime that have impacted as much upon the rural as the urban. There have also been accompanying changes in governing *mentalities*, characterised in terms of a shift away from 'old' criminal justice, and towards the 'new' calculation and management of risk (Garland 2001; Hope and Sparks 2000). The new governmental architecture has been accompanied by a new set of technologies geared towards various forms of statistical crime analysis, and a new managerialism preoccupied with identifying priorities, setting targets and measuring performance, and all of this takes place within the context of a 'preventive turn' (Hughes 2007) in governing crime. However, the extent to which such mentalities and technologies have effected change in the nature of local crime control is still open to question.

In this chapter I will first outline the new architecture of rural crime governance and its associated mentalities and technologies. Second, however, I will argue that despite this new architecture there is considerable space within the rural governance of crime for the presence of a number of different imaginaries or discursive representations of 'the rural crime problem'. These continue to have an important influence upon the way that rural crime is governed in practice.

The New Architecture of Rural Crime Governance

It is probably fair to suggest that for a century and a quarter local crime control in rural areas was largely synonymous with the enterprise, or often lack of enterprise, of rural policing (Cain 1973). Over the last few decades, however, the governmental landscape has started to change due to the emergence of a new 'policy stream' variously referred to as crime prevention, community safety or crime reduction. Whilst each of these terms carries subtly different meanings

(Gilling 2007), they collectively capture significant changes in the way crime governance is imagined.

First, they imply that the objective of crime governance, which might be 'order' or 'safety', is something that needs to be 'co-produced'. It is about the combined efforts of the police and other state agencies working *in partnership* with one another; with other private, voluntary and community sector bodies; and with communities themselves. Secondly, they imply, a 'preventive turn' (Hughes 2007) in the way crime governance is to be achieved. It should be more proactive and risk-aware, identifying possible problems before or as they occur, and intervening in ways that prevent them from escalating into major problems.

Many reasons lie behind the emergence of this new approach to crime governance (Gilling 2007). They include the apparent failure of criminal justice to stem the tide of rising crime in the post-war decades; technological changes; but also changes in the relationship between the state and its citizens that have accompanied neo-liberal globalisation. Greater emphasis has been placed upon the responsibility of individuals and the role of the market in crime control, a process referred to as 'responsibilisation' (Garland 1996).

The first main signs of this new approach began to appear in Conservative policy (Gilling 1997). They made various efforts to responsibilise different agencies to join in voluntary crime prevention partnerships with the police, although the focus of policy activity was mostly on urban areas. They also sought to responsibilise communities through schemes such as neighbourhood watch (Yarwood and Edwards 1995; Yarwood forthcoming). Although, in terms of funding and participation from local agencies, there was a stong urban bias to crime prevention partnerships, greater citizen engagement may have occurred in rural areas. Neighbourhood watch schemes proved to be difficult to establish in high crime urban areas, but they took root much more easily in more cohesive communities, some of which were rural. Significantly, the first neighbourhood watch scheme in England appeared in the village of Mollington in Cheshire in 1982 (Anderton 1985).

Whilst neighbourhood watch was something of a policy success, the same was less true of crime prevention partnerships. The Conservatives were reluctant to use statute to compel agencies, particularly local government, to work in partnership as they did not wish to add powers to a local government sector that they were dismantling (see Gilling 1997). They did however see potential for the engagement of the most parochial level of local government, amongst parish and town councils. This fitted in with the vision of community governance that first appeared officially in the 1995 Rural White Paper *Rural England* (see also Yarwood, this volume). In the subsequent 1997 Local Government and Rating Act, parish and town councils were empowered at their discretion to raise funds to be spent specifically upon local crime prevention measures (Gilling and Pierpoint 1999).

New Labour has continued to sustain and support neighbourhood watch, but it has also taken forward the partnership approach by placing it on a statutory footing in the 1998 Crime and Disorder Act that established 376 Crime and

Disorder Reduction Partnerships (CDRPs) in England and Wales. In England, the CDRPs were established at borough or district council level, which meant, in rural areas, a potentially awkward two-tier structure of district-level CDRPs developing strategies in individual collaboration with county councils and other county-level agencies, such as Drug and Alcohol Action Teams (DAATs) and the police. Perhaps unsurprisingly, in several areas district-level CDRPs effectively merged their operations.

CDRPs comprise partnerships of *responsible authorities* (initially the police and local authority, but subsequently expanded to include fire authorities and primary care trusts), that until recently were obliged to develop three-yearly crime reduction strategies, based upon extensive audits of crime, and consultation with other bodies. This consultation was supposed to extend the partnerships from these responsible authorities to all other local bodies with a potential interest and role to play in local crime control resulting in a locally 'joined-up' approach to crime control.

Parish and town councils had a potentially important role (Lawtey and Deane 2001). Firstly, they were explicitly identified as one of the bodies with which CDRPs would be expected to consult prior to producing their local three-yearly strategies. Secondly, however, Section 17 of the Crime and Disorder Act applied a responsibility to *all* local authorities, including parish and town councils, routinely to consider the crime and disorder implications of *all* of their decisions, rather in the fashion of an environmental impact statement. This was therefore an addition to the power granted by the 1997 Local Government and Rating Act, and it effectively ratcheted up the pressure on parish and town councils to 'do their bit' with regard to local crime control.

Under the 2006 Police and Justice Act CDRPs are now expected to be more 'continuous' in their approach. Thus, in place of triennial crime audits and strategies, CDRPs must now engage in regular 'strategic assessments' utilising the same National Intelligence Model (NIM) that serves as the business model for local policing, and the resultant crime strategies now take the form of annually-updated three-year rolling plans. More significantly for rural areas, and perhaps predictably given the awkwardness of the two-tier structure identified above, the strategic decision-making of CRDPs has been migrated upwards to the county level to be embedded within the 'safer and stronger' themed area of new Local Area Agreements (LAAs) that are the responsibility of county-level Local Strategic Partnerships (LSPs).

CDRPs still exist within district council areas but strategic decision-making is contained in the LAA. The emphasis at the district level is therefore more upon delivery than strategy. Such a change was justified in terms of economies of scale and the reduction of duplication: the same rationale that was employed in some areas to justify the merger of neighbouring CDRPs. The role of parish and town councils remains essentially unchanged by these developments (Gilling 2007).

As noted above, this new architecture for governing crime has been accompanied, in theory, by a new risk-oriented mentality that requires responsible authorities effectively to scan the local horizon looking for potential crime-related problems. Such a requirement has spawned a veritable industry of risk calculation and assessment, including crime analysts; the regular mapping and 'hot-spotting' of crime-related incidents using geographic information systems; the regular provision of crime-related survey data (through the Local Government Place Survey) and police crime statistics, to facilitate the comparison of trends over time, the benchmarking of local performance against the performance of other similar authorities, and so forth. This industry meets the needs of those responsible for governing local crime, but also of those governing the governors: running parallel to the risk calculation and assessment industry is a performance management industry, running from the Home Office, down through regional government offices, to the LSPs and local CDRPs, to ensure that local performance and practice remains on target with national priorities (Gilling 2007).

It would be a mistake to assume that the mainstay of rural crime governance, namely policing, has remained the same whilst this new architecture has been assembled. Policing has been subjected to the same sorts of pressures that have spawned this architecture, and unsurprisingly therefore rural policing has also been subjected to a considerable degree of change. Much of this has been about getting the police service to engage with this architecture: thus, for example, local policing boundaries have tended to become more closely aligned to local authority boundaries, and police managers are much more used to working in partnership with other local agencies, through CDRPs and LSPs. Policing has also become more citizen-focused, as early experiments with community policing have given way to a national programme of neighbourhood policing. This is oriented much more towards engagement with citizens in order to raise public confidence and satisfaction. It is also dependent in part upon the contribution of a new tier of policing, namely police community support officers (PCSOs) as an important part of what is now conceived as the *extended policing family*. Finally, the 'business model' of the NIM represents a concerted effort to shift the culture of local policing away from its 'collar feeling' proclivity, and towards a more risk-orientated and intelligence-led 'problem solving' approach (Maguire and John 2006).

In summary, then, the landscape for the rural governance of crime is much changed. It is more heavily populated, because the number of bodies with a role to play has been increased as a result of the shift in the direction of community governance; and it is busier, because these bodies have been endowed with a set of responsibilities and associated technologies that afford them with various opportunities to govern crime and it is also, therefore, more complex.

Rural Crime Governance in Practice: From Risk to Representation

A question remains, however, over whether and how this architecture is activated in rural areas. The risk-orientation that animates the new governance of crime arguably serves to disqualify or exclude the countryside from serious governmental attention. This, in turn, is because risk-orientated approaches operate with *thresholds* of 'riskiness' that rural crime issues routinely fail to meet.

This problem is further exacerbated by the performance management regimes through which local crime governance is itself governed. A national performance management regime for crime reduction has been in place since 1999, following the launch of the Crime Reduction Strategy and the imposition of a number of Public Service Agreement (PSA) targets and Best Value Performance Indicators (BVPIs) (Gilling 2007; Hope 2005). This performance management regime, inspired by the political imperative of being seen to be 'in control' of the crime problem, has until recently prioritised the reduction of 'headline' crime as the main performance criterion. In practice, this has meant prioritising areas where such crime problems are at their worst, or most spatially concentrated.

At the aggregate national level the poorest performing – i.e. highest crime – areas are all urban, and it is these areas that attract the most governmental attention and resources: the Street Crime Initiative (Gilling 2007) is a good example. Beyond the aggregate level, there is some effort to compare like with like, and this has entailed organising CDRPs and BCUs into 'family' or 'most similar area' groupings, so that performance can then be measured within family league tables. However, the level of resolution employed within this sorting is still wide, and whilst it allows us to scrutinise the performance of predominantly rural areas (as does the Home Office's sorting of urban and rural police force areas (Aust and Simmonds 2002)), it remains the case that within such areas the governmental priority is still afforded to the most urbanised spaces. In the dominantly rural county of Cornwall, market towns were identified as accounting for the lion's share of the crime problem, whilst the more outlying rural areas failed to register as risks (Mawby 2007).

There is, then, a spatial sorting process that serves to prioritise urbanised areas. At the aggregate level, the 'performance' of the government in crime reduction terms depends upon the performance of a relatively small number of high crime urban areas resulting in the trivialisation or marginalisation of rural crime. One possible response for rural areas is to try to compete with this urban bias on the latter's terms, by identifying the kinds of 'urban headline crime problems' that exist in rural areas, in order to attract governmental resources. To some extent this is what they were encouraged to do in their three-year audits, which led to them drawing attention to their most concentrated crime 'hotspots', in market towns. The risk of this, however, is that rural concerns that do not figure on central government's radar are effectively marginalised and thus not addressed.

The result of policy developments over the last decade or so is the establishment of a governmental architecture for rural crime, but, because of the disqualifying effects of its risk-orientation, the absence of clear governmental rationale. There are, in other words, a number of governmental actors present to 'do something' about rural crime, but with no clear steer on what it is that they are supposed to do. This creates a vacuum which is most readily filled by discursive representations of rural crime that serve to animate understandings of what the rural crime problem is and of how it should be governed.

Discursive representations of crime circulate because as Girling *et al.* (2000: 170) observe,

> ... crime works in everyday life as a cultural theme and token of political exchange ... [I]t serves to condense, and make intelligible, a variety of more difficult-to-grasp troubles and insecurities – something that tends to blur the boundaries between worries about crime and other kinds of anxiety and concern.

This 'crime talk' circulates around the networks of actors with a responsibility for the rural governance of crime; a mixture of 'professionals', such as the police, and 'amateurs' such as parish councillors. Rather than simply being a lens through which to view popular anxieties about 'the condition of England' (Girling *et al.* 2000), the crime talk of governmental actors is a lens through which to understand their motives, interests and agendas, and thus to contribute to an understanding of the shape of the rural governance of crime.

Different Representations of 'The Rural Crime Problem'

It is possible to tease out, tentatively, a number of different discursive representations of crime in rural areas. In practice, these may be entwined, and they can be in conflict. What follows, then, is an ideal typology that serves as a useful heuristic device for making sense of the different crime talk of those involved in the business of governing rural crime.

The Idyllic Countryside

The rural idyll features so frequently in representations of the countryside that it would be a surprise *not* to find it here, even though it has become something of a cliché. The idyll paints a picture of the countryside as an orderly and crime-free paradise. As such it is a normative rather than an empirical description that may be associated with the commodification of the countryside (Cloke and Goodwin 1992), since the idyll endows the commodity with much of its value. The idyll is particularly perpetuated by what Woods (2004) calls 'aspirational ruralists', who have often made an emotional as well as a financial investment in their stake in the countryside, and it may be associated particularly with leisure and tourist locations that are dependent upon the promulgation of that image for their continued competitiveness.

With regard to crime, the idyll can work in two different ways. Firstly, it can operate as a discourse of denial, a way of not seeing what is there. The idyll holds on to an impression of the countryside as so self-evidently tranquil that nothing possibly could challenge it. In this regard it is similar to the way that the family is sometimes held up as 'a haven in a heartless world' (Lasch 1977), despite the fact that it so often is not.

Yarwood and Cozens (2004) note that senior police managers sometimes deploy this idyllic representation of the countryside as essentially tranquil and crime-free, and cynically this serves the purpose of justifying the very limited allocation of policing resources to some rural areas. Similarly, Anderson (1999) notes that rural police officers are acutely aware of the limited back up available to them in emergency situations, and it is quite plausible as a result that discretion forms the better part of valour: the rural idyll, in such a scenario, provides a very convenient blinker, enabling rural officers not to see what may be in front of their very eyes. This, however, is not necessarily why front-line police officers may deploy the idyll. It has more to do with managing the potentially excessive demands that can be made of 'street-level bureaucrats' (Lipsky 1980). Street-level bureaucrats deploy various coping mechanisms to deal with the work pressures that they encounter or anticipate, and it may be that the deployment of the idyll works in this way, reflecting back to rural residents what they want to see, whilst simultaneously legitimising the limited attention that they can devote, particularly to outlying rural areas.

The Endangered Countryside

The second way in which the idyll works with regard to crime is by exaggerating threats to it, particularly through a process of 'othering' (Sibley 1995). The representation here is of a naturally crime-free social space that is potentially threatened by criminal others, often outsiders. In this way the idyllic countryside in enlisted in another representation, namely that of the endangered countryside.

As noted above, the threat posed by 'outsiders' is a particularly strong motif in the representation of the endangered countryside. Outsiders include, illegal ravers, gypsies and travellers (Sibley 1995) that are represented as an on-going threat. The idea of 'travelling criminals' also figures prominently, fuelled by media reportage, which often implies that the very success of crime prevention and community safety initiatives in urban areas has displaced 'travelling criminals' into the countryside, in search of rich pickings from country houses and the like.

Although Home Office research evidence (Wiles and Costello 2000) suggested that there has been no noticeable increase in the phenomenon, it also confirmed the fact that crimes in rural areas were committed by offenders who travelled further than for urban-based crime (see also Mawby 2007). Rural police forces often add to this representation by staging frequent 'crackdowns' on major transit routes, although much of this is encouraged less by evidence of a problem than by newly available number plate recognition technology, which allows the police to identify known offenders when they are 'out of place' – a literal but potentially misleading interpretation of 'travelling criminals' that allows the police to present themselves on such occasions as holding the thin blue line against an external enemy.

The Frightened Countryside

In the wake of the Tony Martin case and the politicisation of the countryside (Woods 2004), the Home Office and other governmental agents have started to take more of an interest in rural crime. Analyses of crime statistics (Aust and Simmonds 2002) continue to demonstrate that rural crime rates are considerably lower than crime rates in urban areas. Such comparisons suggest that the sense of endangerment is unfounded but Home Office researchers now stop well short of suggesting as much, perhaps mindful of the controversy that greeted the published results of the first British Crime Survey (Hough and Mayhew 1983), which utilised statistical average risks of victimisation to imply, somewhat condescendingly, that certain groups were a good deal more fearful of crime than they should be.

The impression that crime is a growing problem in the countryside does not simply go away, however. Research into the fear of crime has shown that while the fear of crime in rural areas is less than it is in urban areas, it is nevertheless rising at a faster rate (Countryside Agency 2004); whilst a survey conducted by NFU Mutual found that 82% of respondents admitted to worrying more about crime than they did a year ago (Countryside Online 2004). Yarwood and Cozens (2004), meanwhile, note the tendency of some rural police managers to see the problem in rural areas as one more of fear than victimisation, prompted in part by a sense of isolation, and by the fear engendering qualities of village gossip. More recently, the National Policing Improvement Agency's (NPIA) guide on neighbourhood policing in rural communities notes that '(d)espite consistent falls in the overall level of crime in the last 10 years, many people within rural communities believe that crime is rising and feel increasingly insecure or isolated within their neighbourhoods' (NPIA 2008: 5).

Aust and Simmonds (2002) suggest that there may be a rational kernel to this growing fear as, whilst rural crime rates are significantly lower than corresponding urban ones, there has been a very much sharper rate of increase in headline rural crimes. For example, rural burglary rates in 2001 were three times higher than they had been in the early 1980s, whereas in non-rural areas, after some significant increase up until the 1990s, burglary rates had actually fallen back to early 1980s levels. The authors suggested that this rate of increase 'may help to explain why crime has become more prominent in the agenda of rural concerns' (Aust and Simmons 2002: 1). And they go on to suggest that '(i)t is therefore not surprising that people in rural areas perceived over that period a change in their idyllic environment once relatively undisturbed by crime.'

Overall, what emerges here is a representation of the frightened countryside, in effect saying that the countryside is not as endangered as people might think. This is not the same as the discourse of denial that emanates from the rural idyll, because the fear of crime is not now regarded as something to be taken lightly. The representation of the frightened countryside, consequently, may be accompanied by a response suggesting that what is needed is reassurance, which is to a large extent what the neighbourhood policing programme is about (Quinton and Morris 2008).

Neighbourhood policing, and its aspiration to make the police more visible and accessible is particularly well suited to addressing the representation of the frightened countryside, which, as Yarwood and Cozens (2004) note, plays upon the themes of remoteness and isolation. For the police, promoting the case for reassurance is simultaneously promoting the case for more resources, which may then be used in something of a flag-flying fashion. Arguably the allocation of £60 million in additional policing resources for a rural policing initiative between 2000/01 and 2003/04, as part of the Police Modernisation Fund, had the same reassurance function, addressing not only the apparent fear of crime, but also the perception that rural policing provision was on the wane (Neyroud 2004).

The Deprived Countryside

The representations discussed above tend to offer a view of rural crime from the perspective of the potential victim. There is, however, an alternative representation, of the deprived countryside, that offers a view from the perspective of the potential victim *and* the potential offender. It also reflects the two main competing paradigms of *situational* crime prevention, which is orientated towards the protection of victims, and *social* crime prevention, which is orientated towards the support of potential offenders (Gilling 1997).

The representation of the deprived countryside exploits a discourse of unmet social need, which problematises the issue of service deprivation. Empirical studies of rural public service provision reveal, in comparison to their urban other, a much more limited, patchy and dispersed set of public services, from post offices and transport networks, through to health and social care, leisure, youth service, and of course criminal justice services, symbolised in particular by the closure of many rural police stations in the last few decades.

Service deprivation can be portrayed as enhancing the vulnerability of potential victims, and it is a representation that can be used to add support to the representations of the frightened and the endangered countryside. As Yarwood and Cozens (2004) observe, this image of vulnerability is one that has been played upon by police managers, and used in campaigning for more police resources in areas where remoteness, isolation and lax security can be regarded as problematic. To some extent the lax security is an unintended consequence of the rural idyll, for which open front doors and unlocked cars act as emblematic symbols. Service deprivation can also have very real consequences for certain social groups: Little *et al.* (2005), for example, draw attention to the problems experienced by victims of domestic violence, given the lack of refuge provision, and given the social stigma inherent in admitting to the status of victim, which is also both a consequence and cost of maintaining an idyllic representation of rural life. Victims of rural racism, including Gypsies and Travellers, face similar difficulties (Chakraborti and Garland 2005).

In the case of young people, however, deprivation is often brought to the fore and used as a counter-representation to the one of endangerment in which youths can figure prominently (Muncie 2004). It may be put forward as an explanation for the risky activities of young people that cause annoyance to others and that can bring them to the attention of the police. The main focus here is on the misuse of drink and drugs that have become increasingly available in rural areas (Henderson 1998a). Young people may also be portrayed as being 'stuck' in the countryside, unable to obtain decent employment or housing. They remain dependent upon their families in a way that delays the transition to adulthood, and thus expands the period of time when they are at greater risk of delinquency, because the transition to adulthood is also associated with desistance (Rutherford 2002a). From such a point of view, therefore, addressing rural crime becomes synonymous with meeting young people's welfare needs more effectively.

Discussion

The architecture for governing rural crime has expanded significantly from the days when this was largely the responsibility of the police. When applied to rural crime, however, this architecture struggles because in the rationalistic, managerial terms in which crime reduction is conducted, rural areas tend to be discounted as problem areas. This leaves a vacuum that, it is contended, is filled by discursive representations such as those identified above. The intention has not been to suggest that these representations can be matched to particular interests because representational politics are more fluid than that.

Rather, these representations rather provide different ways of perceiving 'the crime problem' in rural areas, and it is probable that in different situations they are drawn upon by different interests, for different reasons. It may be that particular representations are associated more with particular groups: we can see, for example, how Woods' (2004) 'aspirational ruralists' may have an interest in utilising the representation of the endangered countryside as a way of protecting their own financial and emotional investments; or how rural police officers may prefer to imagine a frightened countryside to which they can administer a reassuring but relatively undemanding presence from time to time. Public services including the police and welfare agencies, meanwhile, can play on the representation of the deprived countryside to lobby for more resources for themselves, and for client groups such as young people who, in the absence of such services, may be more likely to drift into crime.

Precisely how different representations match up to different interests, and how the politics between them are resolved, are questions that demand further empirical enquiry. It is likely to involve different views with regard to the *ends* of rural crime governance, for example contrasting the 'community defence' (Hope 1995) model of situational crime prevention with the more progressive 'pan-hazard' approach of community safety (Hughes 2007). It will also involve contestation: the conflict, for example, between the idyllic expectations of rural residents and the interests of local businesses where a certain level of disorder may be an inevitable consequence of the activities, from the village pub to the summer festival, that they sponsor or support (Barton and James 2003).

It is also possible that other representations might emerge or come to play a more important part in the governance of rural crime. For example, there is a considerable array of social conflicts within the countryside, particularly over access to and use of the land. There are also intra-community conflicts, such as those between young and older people; between white people and black and minority ethnic groups (Garland and Chakraborti 2004), and those within small but significant enclaves of social housing who, in the absence of a formal policing presence, engage in their own 'governance from below' (Stenson and Watt 1999).

In many ways it is remarkable that these conflicts, some of which have at times been quite violent, have *not* merged together into a general representation of the divided countryside, and this possibly bears testimony to the 'silencing' power of other representations. Whilst social divisions do come to wider attention, they tend to be dispersed into single non-crime-related issues, rather than gathered together into the representation of a fundamental crisis in rural society. In so far as they impact upon those with a responsibility for governing crime, they are perceived more as issues of *order* than of *crime*, although such a distinction is a difficult one to maintain. The absence of such a representation draws attention to the importance of non-decision making (Lukes 1976), which helps to keep issues *off* the agenda, and one can imagine how certain interests might coalesce around *not* making a problem out of general social conflict in the countryside.

When rural crime was the main preserve of the police, it was perhaps a little easier to maintain control over a dominant representation of the countryside as a relatively safe place. However, the shift in the direction of community governance has weakened this control, leading to a contested politics over the nature of rural crime, and how best to address it. Such contestation may be played out in a variety of arenas, such as in parish and town council meetings; in neighbourhood policing PACT (police and communities together) meetings; in police and CDRP tasking and coordination groups; in LAA strategies; in the local media, and so on. These are all potentially fruitful avenues for future empirical research to map the rural governance of crime, and to begin identifying which representations prevail in which different geo-historical contexts.

Chapter 8

Big Brother Goes to the Countryside: CCTV Surveillance in Rural Towns

Craig Johnstone

Introduction

Closed circuit television (CCTV) surveillance is probably not something most people would associate with rural spaces and place. After all, even in *1984*, George Orwell's vision of the ultimate surveillance society, the countryside is the one place of escape from the watchful eye of Big Brother. Nevertheless, the UK CCTV boom of the late 1990s and early 2000s (see Gill and Spriggs 2005; Graham 1999; Johnstone 2002; Norris and Armstrong 1999; Webster 2009; Williams *et al.* 2000 for discussion) saw considerable sums of public money spent on bringing video surveillance to the public spaces of provincial towns and even villages across the country. As the *New York Times Magazine* (2001) wryly observed, 'each hamlet and fen in the British countryside wanted its own CCTV surveillance system, even when the most serious threat to public safety was coming from mad cows'.

This chapter begins by tracing some of the specific practical challenges that have accompanied the migration of surveillance technology from the city to the provinces. It then seeks to uncover why CCTV has been introduced to towns and villages which have relatively low rates of crime and, therefore, little apparent need for such technology. Throughout the chapter reference is made to my own empirical research into CCTV in rural areas, which focused mainly on Wales, involving interviews and discussions with stakeholders in rural video surveillance schemes, as well as a survey of public opinion. For reasons of confidentiality, the case study towns and interviewees are not referred to by name and research participants' roles only loosely defined.

The focus of the chapter is rural video surveillance schemes developed by public authorities to keep watch over the streets and public spaces of town and village centres designed expressly to aid in the fight against crime and disorder[1] –

---

1   CCTV is used widely in private leisure and retail spaces but it is deployed unevenly (some shops utilise highly advanced systems while others have none) and there is so much variation in the types of system deployed and how they are used that it is difficult to identify commonalities. Most CCTV research has concentrated on public space video surveillance since this is typically state funded and most controversial due to concerns about the invasion of privacy and the erosion of civil liberties.

the official *raison d'être* of CCTV (see Williams *et al.* 2000). Moreover, although there has been considerable CCTV proliferation in the rest of Europe and beyond during recent years (Hempel and Töpfer 2004; Koskela 2000; Walby 2005; Wilson and Sutton 2004), this chapter concentrates on developments in Britain, where surveillance coverage of rural settlements is already considerable.

The Practicalities of Rural CCTV

The largest big city CCTV schemes can involve the linking of over 200 cameras to large control rooms, containing banks of TV monitors, constantly staffed by teams of operatives. The contrast between such schemes and what is possible in rural areas is often stark. Resource constraints mean that stakeholders in most rural CCTV schemes have inevitably had to make compromises or invest in costly innovative solutions to overcome the diseconomies of scale inherent in providing surveillance coverage for small towns. Rural towns tend to have relatively small zones of commercial, retail and leisure activity. As it is typically only these zones which are thought to warrant CCTV coverage, it is has been hard for scheme proponents to justify the grandiose multi-camera schemes they might like. Indeed, in the smallest towns to have installed video surveillance, cameras number as few as two and seldom more than six.

The implications of limited coverage are twofold. First, the fields of vision of different cameras may not intersect, which means incidents can occur that are simply not spotted or may move out of the zone of surveillance at a crucial moment. Second, it is hard to justify setting up an expensive control room staffed constantly in order to monitor the feed from a handful of cameras watching over a town where crime and disorder is relatively rare. This problem is more complex and has been resolved in different ways in different locations. The cheapest and most simple solution is to provide no formal monitoring of live images at all. In such cases what cameras witness is recorded and may have a role to play in the investigation of offences that by happenstance occur within their fields of vision, but it is the deterrent effect that is prioritised. Although this is the least optimal solution, operating costs of all schemes, even when installation is paid for by government grant, must typically be covered locally so it is sometimes the only option. The scheme set up by the parish council in the village of Lenham in south east England in 1996, whose monitoring station was originally located in the laundry room of a local hotel (*Daily Telegraph* 1997), operated in this fashion for a number of years. Not only does the absence of regular staffing mean that an appropriate response to incidents cannot be mobilised when they occur but any after-the-event searching of tapes for evidence also had to be carried out elsewhere.

A partial solution to the monitoring problem is the use of volunteers. However, it is difficult to tie volunteers to a roster and to find people willing to work at unsociable times when monitoring can often be most useful. Furthermore, my research in rural Wales indicated that local residents grouped such volunteers with 'busybodies' and 'nosy neighbours' and indicated that they wanted the police to monitor the CCTV images. In many rural locations the police are indeed responsible for monitoring, although it is only possible when CCTV control rooms are located in police stations. Moreover, watching the feed from surveillance cameras is typically low on the list of police officers' priorities unless their attention is drawn to an unfolding incident, they are looking for known individuals or their shift is particularly quiet. It should also be remembered that many rural police stations close or are not continuously staffed at night so monitoring is at best sporadic.

As there is often no straightforward way to monitor CCTV pictures in individual rural towns, a solution which is becoming increasingly popular as telecommunications technology develops is to link cameras in a number of small towns to a single control room. This generates economies of scale and makes the employment of specialist monitoring staff viable. A pioneer of this approach was Breckland District Council in eastern England. Its 62-camera scheme provides coverage of five widely spread small towns monitored 24/7 from a control room at the council headquarters in Thetford. Other local authorities including neighbouring North Norfolk have developed similarly elaborate schemes whilst elsewhere new cameras erected in small communities have been linked to pre-existing and already staffed control rooms in larger towns. In west Wales, for example, Lampeter's cameras are monitored 30 miles away in Aberystwyth. But remote monitoring, at least until recent advances in internet-based data transmission, has come at considerable cost, since laying fibre optic cables over large distances to link up each network is extremely expensive. For example, capital costs of the Breckland Scheme were in the region of £1.1million.

For rural police forces the decision about how a system is to be monitored has implications for their resources, although no solution is resource neutral. The mounting pressure placed on the police by technology-driven information flows is well documented (Chan 2003) and CCTV can compound this, especially when it falls to police officers to monitor feed from surveillance cameras or search video recordings retrospectively for evidence of reported incidents. But the monitoring of surveillance footage by specialist staff often only changes, rather than reduces, demands on police time, with requests to search for evidence after the event being replaced by additional callouts to unfolding incidents. Since police officers are not always based in the most rural towns or villages they may have to travel considerable distances to incidents witnessed by CCTV. The reality of rural policing is that a small number of officers must cover very large geographic areas and no amount of video surveillance coverage can change this equation.

The Purpose of Rural CCTV

So far this chapter has identified that there are a number of diseconomies and practical difficulties associated with establishing and operating CCTV schemes in rural areas. Add to these the fact that many rural areas experience comparatively little crime and there seems no clear-cut rationale for taking this technology into country towns and villages. Thus, drawing on interviews conducted with stakeholders involved in CCTV schemes mostly in rural Wales, this section explores the claimed purposes of rural video surveillance.

Preventing Crime

CCTV systems in the towns studied were designed so that they would offer surveillance of the main commercial and leisure streets in an attempt to discourage the most common problems faced: property crime and shoplifting. But even these most common offences were not particularly prevalent and as a consequence few stakeholders expected crime to be driven down dramatically. However, rural areas tend to have long histories of very low crime rates so relatively small increases can seem significant – and in percentage terms often are significant. A crucial role of CCTV, therefore, was to maintain the status quo. As one police officer stated:

> CCTV doesn't relate to the crime rate. It relates to the need to protect people and keep crime low. It's a preventive measure ... More than anything else it is there to prevent problems. (Interview 2)

Although their size means opportunities for crime can be quite limited in country settlements, the crime prevention measures evident may also be less extensive or advanced than those adopted in more crime-prone locations. In addition, rural police forces are unlikely to be able to guarantee the response time to serious incidents that would be expected in urban areas and it is rarely possible for small rural businesses to mitigate the thinness of this 'blue line' by calling on the services of the private security personnel now commonplace in larger towns and cities.

In my case study towns, the imagined origin of the offender against whom CCTV was to be a defence seems to have varied depending on type of offence. Whereas those concerned about disorder and anti-social behaviour (ASB) believed most offenders to be local people, others thought it was outsiders who posed the greatest threat. This 'stranger danger' is further emphasised in the following justification for the introduction of CCTV offered by a senior police officer:

> Something like 40% of all our crime [in the force area] is committed by criminals travelling in from outside ... So the only way we can protect our community is to say to travelling criminals, 'If you come to this area you're going to be caught' ... That means you have to attack the problem in a variety of ways. You have to ... protect your potential target areas, many of which are your town centres. If you didn't have that protection you'd be raising the vulnerability of your community; you'd be saying to criminals in South Wales and Merseyside: 'Here's an easy target, come into our area there are no CCTV cameras you are never likely to be stopped.' We'd be putting our public at considerable risk if we did that. (Interview 2)

While the threat posed by non-locals to the case study towns is debatable, the argument resonated strongly with local stakeholders and has been a significant driving force behind CCTV elsewhere.[2] As one interviewee observed:

> While the crime rate is relatively low here we want to keep it like that. There was a recognition from us [in the town council] that towns around us were starting to [install CCTV] and a general feeling that if we didn't keep up we could end up being the place people go because there aren't cameras and you can get away with it. Crime is low but you can't be complacent. (Interview 14)

As the only towns of significant size without CCTV in the area, there was widespread local concern that they could well become the target for organised crime that video surveillance had driven from other towns, particularly those in the more heavily urbanised parts of South Wales. Crime displacement is a complex process that can be spatial, temporal, or in the type of offence committed (see Barr and Pease 1992) and is argued by critics to be an inevitable consequence of the uneven distribution of situational crime prevention measures such as CCTV (see Crawford 1998; Short and Ditton 1998). A vignette recounted by a police officer suggests such concerns were not wholly unfounded:

> If you go back to before we had the cameras, we had three smash-and-grabs in [an electronics store]. They were committed within a 10-day period by the same team from Pontypridd, and they said the reason they came down and did [the electronics store] and a sports shop in those 10 days was that they were aware that there was no CCTV in our town and they knew it was that much easier to get away with it. So, when they came back two years later and they did a sports shop – they did that sports shop three times – we were able to focus cameras on it during the last occasion and they've been apprehended. They've now made it clear that they're not coming back here because they know CCTV exists. (Interview 3)

Interviewees expressed little concern about the possibility of CCTV displacing crime within the towns and/or further down the urban hierarchy. Most dismissed this as unlikely, unsure what worthwhile alternative targets were left for the organised criminal.

Discouraging Anti-social Behaviour

When crime rates are uncommonly low, it does not necessarily follow that residents will automatically feel safe. Indeed, countless studies (see, for example, Brown

---

2   Its accessibility from the motorway network and proximity to large centres of population made Lenham a popular target for travelling criminals and it was a spate of organised and brazen shop robberies which precipitated demands for CCTV.

1998; Koskela and Pain 2000; Valentine 1989; Young 1988) have shown that the fear of crime and victimisation has multiple catalysts, many of which have nothing to do with the local crime situation. A particular cause for concern in my Welsh case study towns was drink-induced 'rowdyism' (as the funding bids described ASB), especially on Friday and Saturday nights and typically involving the 16–25 age group. This was presented as almost as serious a problem as crime itself and the need to keep it under control an important justification for CCTV installation. As a town councillor commented:

> If you look at the typical crime in [the town centre] away from shopping hours, the main contribution is the usual small town Friday and Saturday night mayhem, whatever causes it, and I think that it was felt that CCTV would have a calming effect on it. (Interview 14)

Recent years have witnessed enhanced focus in government rhetoric and policy instruments on ASB (Millie 2009; Squires 2006, 2008), especially that associated with the night-time economy (Hadfield 2006; Hobbs *et al.* 2005). Crucially, like many rural service centres, the towns researched are popular destinations for a night out, drawing custom from large hinterlands. While the behaviour linked to pub culture does sometimes have a criminal dimension – criminal damage, public order offences and various types of assault are frequently drink related – it was also viewed as an important catalyst for fear, its impact compounded by the fact that many people still reside in and around the central streets where drunken behaviour is played out. The impact of disorder on public opinion was outlined by a town councillor:

> Somebody like myself – who deals with figures and attends police meetings – is aware that we are very fortunate in this town. But it doesn't change people's perception of crime and the fear of crime. They think it's bloody horrendous … but, in fact, there isn't a [crime] problem here. (Interview 8)

According to one police officer interviewed, even the aftermath of a night of drinking – large quantities of litter, broken bottles and vomit on the streets – serve to convince older members of the community that the modern world is a much different and more dangerous one to that in which they grew up.

It is unclear though to what extent CCTV is able to have a calming effect on ASB associated with alcohol. First, drunks are only identifiable if they enter the area watched by cameras inebriated to the point where they exhibit the physical side-effects of drink. But, since town centres are places that people often visit specifically to go drinking, the large majority do not arrive in this state. Thus, the most respectable-looking businessperson has the potential to be a loud and aggressive drunkard. The visual categorisation on which surveillance-based prevention is predicated (Fiske 1998) is therefore difficult and early intervention almost impossible. Consequently, while CCTV is useful for keeping watch over crowds of alcohol-fuelled revellers so a quick response to any breakdown in order can be launched (and evidence of the incident recorded), its potential impact on rowdy behaviour would appear rather slight. This is compounded in rural areas by the small number of police and private security personnel on hand to deal with serious incidents.

Secondly, although Short and Ditton (1998) found that those going out looking for trouble or who had been prosecuted for earlier misdemeanours were more aware of the presence of cameras and as a consequence tended to take their disputes elsewhere, spontaneous crimes of violence, such as drunken brawls, are an immediate response to a certain trigger and unlikely to be deterred by CCTV. Indeed, one police officer suggested:

> CCTV can never replace the policeman walking the street. While CCTV can provide evidence of a fight outside [a pub], the presence of policemen meant it simply didn't happen: a drunk might forget about CCTV but he doesn't forget about a policeman he sees standing on the corner. (Interview 3)

Safety

Given the very low crime rate recorded in rural towns, there is a strong argument that CCTV's prime function is reassurance. A senior police officer suggested in interview that it was the process of installing CCTV – being *seen* to be making the towns safer – that really mattered. One of his colleagues commented:

> I don't think the intention was ever to make the town solid with cameras … It was there to create amongst the elderly people [the feeling that] it's safe to come into town: it's a tool for security and it's achieved that. When you think that this [sub]division covers 853 square miles and has three recorded crimes a day, then it's only the perception of the quality of life that you can combat. How can you decrease three crimes per day? You can't, can you? (Interview 3)

There is undoubtedly a certain degree of logic to claims that the installation of CCTV could improve perceptions of safety, but there are also flaws to the argument. The limited geographical coverage provided by the cameras in rural towns, the inability of cameras to reduce the non-criminal but nonetheless intimidatory aspects of ASB, uncertainty as to whether someone is actually watching the images filmed 'as they happen', and the realisation that even if someone is pointing a camera at you being assaulted the police may be 20 minutes away, make it hard to see how video surveillance can reduce real or perceived danger. It is actually conceivable that the presence of cameras could make some people feel more at risk than they did before, as the need for CCTV would seem to indicate that all is not well in what had previously been regarded as safe towns (cf. Gill and Spriggs 2005).

The Politics of Rural CCTV

The weight given to each of the rationales for rural CCTV proliferation articulated so far is likely to vary depending on local circumstance, but informal discussion with stakeholders in schemes elsewhere in the UK suggests these are fairly typical. However, Big Brother did not take up residence in the countryside solely to future-proof small towns and villages against crime and disorder. As this section explains, local and national politics were also crucial (see Williams *et al.* 2000 for additional discussion).

The Politics of Envy

In towns and villages without CCTV, the introduction of cameras to neighbouring settlements, or rival places of comparable size and function, have often stimulated debate about the need to 'catch up' and not be 'left behind'. While fuelled in part by concern about crime displacement, envy also appears a factor. In one of the rural towns researched, the presence of CCTV in the nearest town of comparable size and function was an important catalyst for scheme development, influencing both the desire for CCTV and the form it was argued the system should take. The successes attributed to this rival town's scheme were noted regularly in the local press, the police spoke positively of it and the Home Secretary even made an inspection visit. It is notable that many councillors who became highly supportive of CCTV had made special trips to view this system (and this one only) in operation. As a police officer interviewed observed:

> The success there has been quite remarkable. Having said that, it's had success elsewhere in the UK as well, but this scheme was closer to home. (Interview 3)

And as a county councillor recalled:

> I think I became aware of [the desire for CCTV] when [local] members put it forward at county council level because they'd heard reports on what was happening in [rival town]. (Interview 8)

There can be little doubt that the presence of CCTV in this rival town convinced some people that their town also needed CCTV: the 'you've got it we must have it' mentality. While not a sufficient catalyst on its own, it was an important contributing factor. Significantly, for reasons of cost the system installed in this particular case study town was much less extensive than that boasted by its rival. This did not go un-noticed by local councillors who, within weeks of the scheme's launch, were calling for upgrades and extra cameras.

The Politics of Policing

It is evident that the police have much to gain from CCTV, earlier comments about impact on workload notwithstanding. The camera can define the 'who, what and where' of town centre crimes, saving both operational and administrative resources. The manpower needed to deal with an incident can be estimated before officers are deployed; CCTV identifies the culprit, freeing up the hours that would otherwise need to be spent cross-checking witness statements; and, faced with camera footage, most offenders plead guilty – more time saved (Williams and Johnstone 2000). As such, some senior police officers have gone so far as to liken CCTV cameras to full-time policemen who are on the beat 24 hours a day, all taking notes and never going off sick (Graham *et al.* 1996: 8). It is perhaps no surprise then that my research found the police to have played a key role in stimulating local support for video surveillance schemes amongst key stakeholders.

Members of the town councils and local authority in the case study towns all commented that the original idea had not originated from their colleagues or council officers but from the police:

> I think the idea was planted by the police. (Interview 14)
> The police, if I remember correctly, were saying for some time that it would be
> a good thing to do. (Interview 11)

Even a senior police officer observed:

> I think it could have been the police; I think we started the ball rolling, possibly
> at a seminar that was held at headquarters. (Interview 1)

Moreover, the police certainly made it clear that Home Office grant aid was available and provided a not-to-be-missed opportunity. A senior police officer observed:

> We supported it from day one ... We think it's a good use of technology ... it's
> an extra pair of eyes on the street for us really. (Interview 1)

However, this statement seemed to contradict the official force policy as articulated by the most senior officer interviewed:

> All the schemes have started from the community. We have not had a policy to
> go round and tell people they should have cameras. (Interview 2)

In many respects this statement cannot be contested, as the police never actually told another agency that they *should* introduce cameras. But they had no need to be so blunt, as their more assiduous approach seems to have been effective in convincing decision makers to take CCTV seriously. Their gospel of video surveillance was appealing, and those who were the first to be converted after being shown the rival town's scheme and its successes (by the police) helped to spread the word to others and reinforce their message. This was important, because for CCTV to become a reality the police needed support from a cross-section of agencies since they could not 'go it alone' and fund CCTV from the police budget: it cost too much, the Home Office was not prepared to finance video surveillance initiatives that only the police were involved with, and one officer also observed that, had it been police-led, CCTV may have been perceived as 'a Big Brother thing' (Interview 1). A more senior officer was also keen to emphasise that:

> If [CCTV] was just something that the police were involved in then I think that would be wrong. We are part of the community; this is one way of protecting the community, so it needs to be an involvement and commitment for a wide range of people. (Interview 2)

It seems that the police, although eager to reap the benefits CCTV could offer, simply did not have the financial or political capacity to act and realised they needed to generate a consensus behind CCTV amongst other key local actors for it to become a reality. Highlighting the Home Office grant scheme proved an effective way of stimulating it.

The police were aided greatly by the passing of the Crime and Disorder Act in 1998. This made local authorities statutorily responsible for crime and disorder prevention and required them to work in partnership with the police and other local institutions to achieve this objective. Being involved in a CCTV scheme was evidence of this partnership in action and ensured the ongoing commitment of local government to such initiatives.

The Politics of Funding

Whilst there are a small number of examples of provincial towns and villages that have funded CCTV from existing budgets and locally-raised monies, Kings Lynn and Lenham amongst them, the costs involved are usually beyond rural communities since they do not have access to the sort of public sector budgets and private sector wealth that are regularly tapped in larger urban areas (Bannister *et al.* 1998; Beck and Willis 1995; Ditton *et al.* 1999; Fyfe and Bannister 1996; Short and Ditton 1996). Central government funding therefore made rural CCTV schemes feasible. Moreover, in the schemes researched it was evident that, by making CCTV affordable, the funding competitions put CCTV high on the local agenda, an agenda on which it would probably not at that time have featured despite the best efforts of the police. Attitudes towards the Home Office grant were, therefore, uniform:

Oh, [it was] most definitely a catalyst … If the Home Office hadn't introduced the scheme to offer the money in the first place I don't think we'd have even though about it. (Interview 8)

There's no way we could have persuaded the county council to part with that much money if there hadn't been the grant, I'm absolutely certain of that. (Interview 14)

The only representative out of step with this consensus spoke for the police. A senior officer pointed out that CCTV had been introduced to other towns in the force area without Home Office funding, and argued:

I think the first decision is: Is this something you want to do? The next decision is: how is this going to be funded? And if there is money available from the Home Office then so be it. If not, you have to find some other way of looking at it. I think it will happen because the public want it to happen. (Interview 3)

The suggestion that CCTV partnerships are bottom-up initiatives may still be true of some locations but it is hard to see where funding would have come from for the schemes researched and many others in rural areas if Home Office grants had not been available. In the cases documented, demand *followed* the financial solution. That bids were made for CCTV funding arguably says more about the funding mechanism than it does about the desire to introduce cameras, the police excepted, that existed in these towns at that time. In many respects it was a case of bid while we can, sort out how to best configure, utilise and operate the systems later. The catalytic role played here by central government funding reinforces Healey's (1998) claim that funding is not responsive to local problems: it imposes solutions which may not be the best for the issues facing that specific locale. Indeed, Koch (1998) points out that in the last years of Major's Conservative administration, 78% of the Home Office crime prevention budget was spent on CCTV; rather than just one amongst a suite of crime reduction options offered to local partnerships, video surveillance was often *the only* option.

Conclusion

In spite of the relatively low levels of crime and disorder recorded in rural areas and the practical difficulties associated with realising the full potential of surveillance technology in such locations, public space CCTV cameras are now a well-established feature of British country towns and villages. This chapter has sought to demonstrate that, whilst concerns about crime, especially that displaced from elsewhere, and ASB associated principally with the night-time economy have been mobilised as justification for the proliferation of rural CCTV, other factors have also been crucial. In particular, the important role played by the police in generating support for video surveillance, the significance of rivalries between towns sparking a desire not to be left behind and, critically, the availability of central government funding which enabled many rural communities to think seriously for the first time about installing a CCTV system. Whether these rural schemes will be viewed by history as a useful late 20th century innovation or politically driven folly depends very much on how success is measured. The small number of crimes recorded and the potential for displacement makes it difficult for rural CCTV to be a resounding success purely in crime reduction terms. Consequently, 'success' is more likely to emerge from the use of cameras to aid police deployment and detection and in the important but less tangible (nor directly attributable to CCTV) public perception of safety and fear of crime.

Chapter 9

Whose Blue Line is it Anyway? Community Policing and Partnership Working in Rural Places

Richard Yarwood

Introduction

Many first world countries have adopted neo-liberal policing models that emphasise, on the one hand, closer cooperation between the police and public and, on the other, tighter monitoring and regulation of policing through various systems and measures (Crawford *et al.* 2005; Hughes 2002). These principles have become enshrined in partnership working, which now pervades policing and governance at a range of scales (Crawford 1997; Gilling, this volume).

A key but problematic element of partnership work is to determine responsibility for action. Critics have suggested that poorly organised partnerships can lead to a 'nobody in charge' world where no agency is willing to accept responsibility for managing a problem (Goodwin 1998). Consequently, a discourse of 'responsibilisation' has emerged in recent years (Herbert-Cheshire 2000) that has sought to define who is accountable for what and where. Increasingly, greater emphasis is being placed on local communities to provide local solutions to local problems, including crime and its policing (Yarwood 2007a, 2007b, 2010). Such thinking builds upon the active citizenship policies of the 1990s that encouraged a range of local policing schemes, such as neighbourhood watch, with varying success (Fyfe 1995). The current emphasis on community-based management of crime has been supported by legislation, surveillance and performance monitoring from central authorities to such an extent that it has been suggested that this is leading to a form of governance *through*, rather than by, community (Woods 2006; Woods and Goodwin 2003). In other words, local communities are being manipulated from afar to implement central crime and policing policies.

There is evidence from urban areas that some local communities are unwilling or poorly placed to engage with this work (Hughes 2002); that professional keepers of the law are unwilling to relinquish their roles and powers (Newburn 2003) and that, despite talk of a shift from government to governance, local authorities continue to row, steer and coach partnerships (Phillips 2002). Questions remain, therefore, about who is responsible for local policing in both *de facto* and *de jure* terms.

Although faced with challenges of distance, isolation and fewer resources, partnerships in rural areas have the potential to build upon closer community structures and stronger traditions of voluntary working (New Zealand Police 2006) or, at least, perceptions thereof (Woods 2006). This chapter examines three attempts to encourage partnership working in three different countries, England, Australia and New Zealand, and questions who should be and who is in charge of rural policing.

Rurality poses different challenges and expectations across space. The places presented in this chapter are very different in terms of their relative isolation and how they are imagined and contested. Australia faces challenges of immense distances as well as cultural and political tensions in the ways that Aboriginal peoples have been, and are, policed (Cunneen 2001; Carrington 2007; Yarwood 2007b). The police in rural New Zealand also have to content with isolation but have a strong tradition of working with local communities to overcome this problem. Although rural areas of the UK are more accessible, the countryside is a pressured space that is contested by many different groups of people with different visions of rural life. Cultural as much as criminal threat has influenced the direction of rural policing (Yarwood 2001; Yarwood and Cozens 2004). In all three countries, however, rural society has common characteristics including new populations with diverse lifestyles, economic restructuring and relatively low crime rates (but see Hogg and Carrington 1998, 2006). Further, a plethora of partnerships and community initiatives has been developed in each place, reflecting the operation of neo-liberal restructuring. While some of these, such as Aboriginal patrols, are unique to local circumstances, many others, such as rural neighbourhood watch schemes, are found in all three countries.

The following case studies examine particular initiatives in each country; namely: a rural policing initiative in the West Midlands of the UK; a partnership scheme in Western Australia and community patrols in the central police district of New Zealand's North island. They draw upon interviews and observations with police officers, government workers and members of the public conducted by the author between 2000 and 2008. The chapter examines these schemes with the aim of revealing how policing partnerships are organised, driven and coerced. In turn, these findings are used to consider who is in charge of policing a 'nobody in charge world'.

England

Recent Rural Policing Initiatives in the UK have been driven, in theory at least, by legislation. The 1998 Crime and Disorder Act required local authorities and the police to work in partnership with each other and other agencies to improve community safety and reduce crime. Although the 1998 Crime and Disorder Act has had a significant impact at the county and district level, it failed to coerce local authorities at the parish level to engage meaningfully in crime prevention and policing in rural places.

Of particular significance was a requirement for parish councils to be involved in local crime prevention strategies. Parish councils are the lowest democratic tier in England and are comprise of lay members of the public who are elected or co-opted to serve on them. They have very limited statutory powers, confined to the management of allotments, footpaths and other local open spaces, but are valued for their potential to speak democratically for local interests (Cullingworth and Nadin 2006). Following the publication of New Labour's Rural White Paper in 2000 (DETR 2000), there have been efforts to modernise parish councils by auditing their accountability and building upon their capacity to empower local people (Gardner 2008). Parish councils were afforded two roles under the 1998 Crime and Disorder Act, which reflected this agenda as well as a drive to make local people more responsible for local crime planning.

First, Parish Councils were required to crime-'proof' their work by evaluating the potential impact of their work and decision-making on crime and disorder (Deane and Doran 2002). They were also expected to pay for any measures to reduce crime from their fairly meagre funding streams. Second, Section 17 of the 1998 Act required PC's 'to be fully involved in their local Crime and Disorder Reduction Partnership (led by the district authority and police force)' (Deane and Doran 2002). Here PCs were expected to feed into strategic planning at the district and council level and, in doing so, to influence the direction of policing in their parent authorities.

However, only 21% of parish councils became involved in Crime and Disorder Reduction partnerships and only 5% of PCs said that they had been delegated any responsibility by these partnerships (CRC 2006a; see also Yarwood 2005). This lack of involvement is more remarkable given that over 80% of parish councils said that they had 'discussed crime and disorder in the past three years' (CRC 2006a: no page). Given that PCs had a legal requirement to participate in these partnerships under the 1998 Act, this lack of participation is notable but perhaps not surprising. Three main explanations can be advanced.

First, many councillors were unaware of this legislation. Although the Countryside Agency issued guidelines on the legislation in 1999, only 20% of clerks could recall receiving this in 2006 and only 5% found it useful (CRC 2006a). More significantly, ignorance of the Act points to a significant disjunction between decision making at the county level and local consultation, let alone participation. This not only raises questions about the effectiveness of crime and safety partnerships in rural areas but challenges New Labour's efforts to democratise and improve the efficiency of PCs (Gardner 2008). At the one level such confusion hints at the difficulties of managing the 'nobody in charge' nature of partnership governance; at another it confirms that the burden of auditing and planning crime and disorder has fallen largely on the shoulders of the police and county/district councils (Gilling, this volume; Phillips 2002).

Second, PCs did not feel it was their job to plan policing in their parishes. The CRC were unable to find any parish councils that had used, or planned to use, the powers under the Act (CRC 2006a). In the eyes of many clerks, responsibility for policing continues to lie with the police (Yarwood 2005, 2010). Certainly rural lobby groups, such as the Countryside Alliance, blame a lack of policing on poor funding from central Government and what they perceive as a lack of support for rural areas (Woods, this volume; Yarwood 2008).

Finally, it is also important to remember that rural policing is simply not an issue for many parishes because rates of crime and the fear of crime remain relatively low in rural places (Yarwood 2010). Consequently, as the CRC (2006a) report revealed, few PCs felt a need to act as crime and disorder were simply not issues.

It is possible to conclude that the 1998 Act had little impact on parish councils in rural areas of England. Despite a clear rhetoric of local responsibility (Higgins and Lockie 2002), Parish Councils shrugged off this liability and were content to let the police police rural places. The thin blue line was, however, a little faded and hard to see in some rural places. Consequently, a second initiative aimed to restore its sheen. The Rural Policing Fund (RPF), which operated between 2001 and 2006 in England and Wales, aimed to improve the visibility and accessibility of the police in the countryside (Yarwood 2008).

The scheme led to a plethora of community-based initiatives including hybrid police officers (who had both a community and emergency role); mobile police stations; specialist police officers; closer liaison with other rural services; community liaison groups and bicycle patrols. One example was West Mercia's Rural Policing Initiative (RPI) (see also CRC 2006b), which was an effort to encourage local people to work with the police to identify and tackle crime and disorder in their parish. The officers concerned designed a resource pack that could be used by any beat managers with a rural parish to identify problems. It clearly had potential but was piloted on four villages with very few crime problems. Consequently only one of these saw the project through to its completion, which to its credit empowered citizens to deal with some issues that had been bothering them. This pilot led to the establishment of 13 other schemes in the Constabulary and the development of a training package for use by other forces. However, with the cessation of the national RPF in 2006 and the publication of a National Policing Plan (Home Office 2004) that made no mention of rural policing, these schemes seem to have withered. This also seems likely to be the fate of other RPF initiatives in other places. On a positive note, though, the RPI was used to inform the development of local policing teams required under the National Policing Plan. However, the example suggests that rural policing initiatives have been rather fleeting and liable to be blown away as the winds of policy changed direction.

More generally, the English example shows that the public, police, local and state government have all been offered the baton of responsibility for rural policing and all have attempted to pass it on. On a daily basis, crime concern remains low and, hence, rural policing is not a priority for any of these groups. Only when the next policing crisis comes along, another Tony Martin incident perhaps, will it be seen whether the baton has actually been dropped or who is left holding it when the music stops.

Western Australia (WA)

In 2004 the Western Australian Government rolled out a 'Community Safety and Crime Prevention' (CSCP) programme as part of a state-wide crime prevention and reduction strategy that sought to address key crime concerns in the state (OCP 2004). The CSCP programme was driven by the State Government's Office of Crime Prevention (OCP), an agency tasked with coordinating 'crime prevention initiatives and community safety activities in Western Australia' (OCP 2005). The OCP sought to develop partnerships with local authorities to identify and deal with crime and safety issues at a local level. It was not the first time this approach had been employed in WA but a review of the previous scheme, Safer WA, argued that greater emphasis should be placed on the local authorities, known as Shire Councils, to devise these plans. It was hoped that their involvement would lead to better co-ordination of local agencies, improved opportunities for consultation (especially of minority groups) and more formalised ways of working (OCP 2004).

Shires represent the second tier of government and are often described colloquially as in charge of 'roads, rates and rubbish'. There are 144 Shire Councils across WA that cover a range of localities including suburbs of Perth; small towns and, mainly, a range of rural, pastoral and remote places.[1] Although efforts were made to develop a rural crime prevention strategy (see Yarwood 2007b), the CSCP, which applied to urban and rural shires alike, emerged as the main way of delivering crime prevention in rural areas.

The CSCP strategy placed emphasis on Shire Councils to act as the lead agency in local crime and safety planning. In some respects this mirrored the UK's 2001 Crime and Disorder Act but, unlike the UK, there were no legal requirements for local authorities to participate in the programme. Instead, Shires had the option of signing a formal compact with the OCP. This committed Shires to write a local policing plan based on a 'meaningful consultation' that recognised local crime and safety concerns; devised ways of dealing with them and actioned appropriate agencies to implement these solutions. Before implementation, these plans had to be endorsed by the OCP and were then subject to monitoring, review and evaluation.

In return for signing, the OCP offered the following (Blight 2004):

---

1 Policing in Aboriginal lands is determined by Aboriginal Justice Agreements Shires and so the CSPC programme did not apply to these areas (Yarwood 2007b).

- Expenses. Shires were offered a one off payment of $10,000 to develop their plans together with a $1,200 annual payment to cover administration costs. A further $20,000 was also available towards tackling a specific issue identified by the strategy. This money was only awarded on the signing of a formal accord between the Shire and OCP to develop and implement a plan.
- Expertise. Access to the OCP's operated a community engagement team that offered shires advice and advocacy on developing crime plans. Additionally these teams made crime statistics and other data available to Shire Councils to aid local planning
- Extra Grants. The development of local plans allowed Shires to access various programmed funding (including the Local Government Partnership Fund; Indigenous Partnership Fund; Community Partnership Fund; Research and Development Fund; and Crime Prevention through Environmental Design Fund) to support and implement local government and community crime prevention initiatives. In 2004/5 a total of $5 million were available through these schemes.

In 2003, the OCP started a community engagement programme aimed at encouraging Shire Councils to participate in the scheme. This involved an outreach team from the OCP's central office travelling to Shire Councils to convince them to sign up to the strategy. During the initial round of consultations, efforts were focussed on 30 Shires that were identified as priorities by the OCP. Following this, the remaining local governments were invited to apply to be considered as part of the CSPS programme. Of these, the 50 seen by the OCP as the highest priority were offered funding and support. This prioritisation implied that the scheme was being driven by the State Government from the 'top down'. By 2007, the programme had been rolled out to other Shires and 120 out of 144 had signed a compact with the OCP to deliver a local policing strategy (Anderson and Tresidder 2008).

During 2004 the author was able to travel with OCP officers to a number of meetings between OCP officers and shire Councils. A total of 11 meetings were attended in a range of rural and remote localities across the State including very remote communities in the Gibson desert.

Reaction to the CSCP initiative by the Shires was mixed. Some shires were opposed to or sceptical about the partnerships for the following reasons:

- Cost shifting. Some councillors felt the CSCP was merely shifting cost and responsibility from the state government.
- Changes in government. There was concern that a change in the state government would lead to the abandonment of the initiative and its replacement with another scheme. The Shire's time would be wasted.
- Not their responsibility. Some councillors felt that crime and safety issues were simply not their responsibility. Responsibility for policing lay with the state government and the police.

- Position of other shires. Many councillors were keen to know whether other Shires had signed for the scheme.
- Specific concerns. Some shires noted examples of good practice that they were already engaged with. For example, one employed a crime prevention officer.

In response to these concerns the OCP officers stressed that they valued the local knowledge of the Shire Councils in developing crime prevention schemes; that support was available from the OCP and that funding opportunities were available for participating Shire Councils. Indeed, where support for the CSPC was strong, shire councillors recognised the opportunities offered by the scheme and listed local projects, such as street lighting or employment of staff that could be achieved by the scheme.

However, funding offered by the programme was a divisive issue. On the one hand some councillors felt they 'might as well take the money as it is not coming from our pocket'. Others, though, felt some pressure to sign up for partnerships, with all their obligations and conditions, as they would otherwise loose money. One councillor noted that Shires would also miss out on expertise and government information if they failed to sign up. In a strongly worded message to me she said 'shame on OCP for withholding funds and valuable crime data if we do not sign'.

At an institutional level the police were required to support the CSCP. However, many local officers seemed ambivalent at best to the scheme. In one case an officer also served as a shire councillor and in this role she had voted against signing a compact. The OCP officers went to her senior officer who said that he would put pressure on her to accept the compact. Others voiced concern that the CSPC was moving responsibility away from the police towards shires, which didn't best please them.

Despite these teething troubles, a recent evaluation of the CSCP strategy suggested that the levels of local authority participation were highest in Australia for a crime prevention planning programme and that consequently this had led to 'crime prevention activity being undertaken at both a planning and implementation level throughout WA', including rural and remote areas (Anderson and Tresidder 2008: vi). The same report highlights several examples of good practice and suggested that Shire Councils felt that the plans led to more strategic ways of addressing crime. However, a common concern was that Shires felt isolated once they had developed their crime prevention plan and wanted more contact with the OCP.

The case of the CSPC shows the state government encouraged local communities, through their local state, to become more responsible for their own protection. Communities with the greatest problems (seen as most 'in need' by the State) were those that were targeted first in the programme. Local communities were made responsible for policing through conditional funding streams and expert knowledge that only became available when local shires formally complied with requirements, laid down by state government, to write, monitor and respond to local crime plans. The alternative, that seemed unpalatable to the vast majority of councils, was to lose the financial and organisational support of the State Government's OCP. Rather than withdrawing from decision making, WA's state government continued to influence the direction of local policing 'at a distance' (Garland 1996; Higgins and Lockie 2002). Whether Shire councils liked it or not, the CSPC moved the responsibility for shaping policing policy squarely on their shoulders.

New Zealand

In New Zealand, unlike the UK, local councils and the police are not required to produce local crime and safety strategies (Coquilhat 2008). Similarly, there is no legislative basis for community policing in New Zealand. The only recent legislation that has impacted on partnership working has been the 2002 Local Government Act that has required councils to devise long-term community plans in collaboration with relevant partners and, in some cases, this has led to closer working between police and local authorities. However, for the most part, community-police relations are unregulated at present[2] and there is little policy guidance for the policing of rural places (New Zealand Police 2006).

In 2006 there were 64 one-person police stations in rural places (Hansard 2006). Officers are expected to be 'practical and mature all rounders' (Buttle 2006: 38) who can work with various elements of the rural community, drawing on personal initiative and intra-personal skills. One officer I met in rural New Zealand seemed to fit this brief perfectly. As well as policing, he had bought a farm in the area and was therefore aware of local agricultural issues. He was also chair of the squash club, a school governor, a member of the local rugby club and allowed hunters to use his land. He liked meeting people at these clubs or during his spare time as he felt these were useful sources of information. Links were also maintained via a 'Plod Page' in the local newsletter that gave forthright views of wrongdoers, referring to them as vermin! His commanding officer conceded that the officer was allowed a large degree of self-management as long as he kept his crime figure down. This allowed some flexibility for him to run the farm and for some unconventional policing. He also felt that running drink-driving tests on local people compromised his standing and goodwill in the community. Consequently he drew on traffic cops from a nearby town to undertake this duty[3].

---

2   The New Zealand Police Act is currently being reviewed and the question of closer regulation is being raised (New Zealand Police 2006).

3   In an extreme example, the media have recently reported on the case of an off-duty rural police officer who was prosecuted for driving whilst drunk to help at an accident. The officer was discharged without conviction by a judge who noted 'There was nothing personal in this for you and it reflects a commitment to your job which you should be proud of'. *Taranaki Daily News*, January 19th, 2007.

This times informal and unconventional approach to policing led, he felt, to good community relations and an acceptance by local people that policing was limited in their locality.

Despite this intimacy, police officers are expected to police vast distances; often from one or two officer stations (the officer in question stated that he covered over 1,000 km a week on patrol). If emergencies occur, support can be distant and disconcerting for officer (Buttle 2006) and public alike. There can also be high turnover rates of staff:

> We've had heaps. Some come and might stay for four or five years and then move on to get their promotions. (Interview with Town Mayor in Central District)

> There is a 'Generation Y' who prefer to be stationed in towns where the girls and pubs were. Rural service does not appeal to younger officers. (Inspector A of Central District)

Single officer stations can also be stretched by illness and absence. In some cases temporary sworn officers have been used in rural areas. In common with other emergency services in New Zealand, other voluntary schemes have developed to fill perceived gaps in policing. These include Neighbourhood Support groups and Community Patrols.

Community patrols have been operating in towns and suburbs for a number of years. They involve residents patrolling towns at night in unmarked cars, recording any events they regard as suspicious. This initially occurred informally but, since 2001, Community Patrols have been encouraged to 'affiliate' with Community Patrols of New Zealand (CPNZ), a national umbrella organisation, to regulate standards and share support and advice. Consequently patrols have become more formalised with requirements for members to be vetted, trained and to serve a probation period before gaining full membership. Patrols are also issued with radios by the police for communication with them on designated channels. This is indicative of a close working relationship with the police and, indeed, affiliated patrols are required to maintain formal liaison with the police. Nationally, memorandums of understanding have been signed between the police and CPNZ to recognise mutual benefit and support . Currently over 100 patrols have become affiliated in mainly, but not exclusively, in small rural towns. According to CPNZ (2008), in 2007 Community Patrols covered over 38,000 km in nearly 72,000 hours of patrolling and reported 16,163 incidents (and these records are only based on returns of 45% of groups).

New Zealand community patrols are not unique. Night Patrols, operated by Aboriginal volunteers are often the only regular presence in remote areas of Australia and were partially established out of a traditional mistrust of the police (Blagg 2003). In the UK efforts in the 1980s to encourage residents to set up patrols and 'walk with a purpose' largely floundered as this was seen as the police's job (Fyfe 1995; Yarwood and Edwards 1995). In New Zealand, however, patrols work closely with the police and appear to have been established by grass-roots concern. While recognising that rural towns were under policed, many community patrols were started because some residents have felt a duty to help the police. Two examples illustrate this:

> In the last few years the Grey District has had problems with wilful damage and drunken disorder on our streets especially on Friday and Saturday nights. The police can't be everywhere all the time and due to the varied incidents they find themselves attending on some nights, they don't always get the luxury of patrolling and just keeping an eye on things. (Greymouth Community Patrol Website)

> The reason why started it was that there were a large number of burglaries at the time ... and the business community were getting fed up with it ... We were another set of eyes and ears for the police. We weren't there to do their job but we were there to help them. (Interview with Mayor in Central District)

Funding for the patrols often comes from local businesses. In one example, an insurance company has paid for a support vehicle, equipped with search lights, to patrol one area of town. Most patrollers, however, use their own vehicles and often pay their own fuel costs. Community patrollers argue that their work is effective and has been responsible for reducing local crime rates. Their effectiveness, however, relies on close support from the police and local businesses. On the whole this is forthcoming. Three interviews with different police officers in New Zealand revealed high levels of support:

> Our patrols have been initiated by both the police and the public. They are a good way of getting people involved if they that we [the police] are not doing enough in their area. (Interview with Town Mayor in Central District)

> Community patrols are helpful to us. Fair play to people who are willing to give up Saturday nights and maybe have bottles thrown at them. The only problem from our point of view is that crime rate appears to go up as more gets reported by the patrols. But overall this is a good thing as we can act on it. (Inspector, Central District)

> Community patrols are good things because people are empowered. The local police should use their local patrols because people are willing to help. (Inspector, National Police Headquarters)

Similarly a local mayor reported that also reported that her patrols had high levels of cooperation with the local police sergeant:

It is a personal relationship and we have been very lucky. The senior sergeants we have had here been great. In other places the mayor and the senior sergeant do not get on. (Town Mayor in the Central District)

This cooperation had been forged with regular meetings, updates and invitations to the police to work in local planning partnerships. Unlike the UK where such relationships are required under law, these local relations rely heavily on the initiative and personalities of local leaders, police and the public.

Community patrols, however, are not without tension. They tend to operate in the centres of settlements, including small country towns, with Neighbourhood Support groups keeping watch on residential areas. One Neighbourhood Support coordinator reported some tensions between the groups as they competed for funding and, perhaps, recognition. Such competition, though, can arguably be healthy, indicating a present and committed voluntary sector. There is also some evidence that community patrols only represent some aspects of the community and that other parts of it feel spied on or too closely surveyed by their patrol. Thus the Greymouth website claims:

A community patrol offers them another few sets of eyes and ears only. We are NOT the fun police and we don't spend all night 'narking' on people who are just out having a good time. Our members are normal people who also like to let their hair down and have fun too. The Patrols are about normal people doing what we can to help make our community safer for our children, parents, grandparents, friends and neighbours. We have chosen not to sit back complaining about the problems our community faces with regard to crime, but to step forward and to help where we can. (Greymouth Website)

Nevertheless, community patrols do appear to be successful in reducing anxiety and strengthening police and public relations. Unlike the UK and Western Australia, their success appears to rest not on legislation or the incentive (or threatened withdrawal) of funding but on a 'strong tradition of voluntary work and community service amongst the emergency services' (New Zealand Police 2006: 6), especially in rural places. The mayor alluded to a tradition of voluntary working in rural New Zealand:

New Zealand has been run by volunteers for volunteers ... There is huge list of voluntary organisations. People do because they are interested in it. They have a safe community and know that there is not the funding for it. But they do it because they are passionate about it. (Interview with Town Mayor in the Central District)

Examples of these organisation include rural fire teams, civil defence teams, search and rescue teams, St John's Ambulance and Red Cross teams to name a few. The scale of this effort in rural New Zealand is immense and goes some way to explaining why voluntary policing schemes have attracted members and worked well with local police.

Volunteering is not, however, a panacea. Many organisations, include community patrols, have noted that it is getting hard to attract younger members. One mayor acknowledged that patrols were mainly staffed by retired people but suggested that those who benefited from but did not contribute to community patrols should take more responsibility.

> We haven't got huge numbers helping but we'll see numbers growing as it has been pointed out to the business community that there are people out there who have not got businesses and we feel that it is time they took a turn. (Town Mayor in the Central District)

At a strategic level, questions are being asked about the sustainability of voluntary schemes and whether, as in the UK and Australia, these relationships need to be formalised. There has already been, for example, a memo of understanding signed between voluntary search and rescue teams in New Zealand and the police. With this has come the development and formalisation of national standards for search volunteers. The New Zealand tradition of volunteering appears to be moving towards formalisation and surveillance. It is only hoped some of the intangible community responsibility that has served NZ so well, so long is not lost these changes.

Conclusions

Through the examination of three different efforts to encourage community-based policing in rural places, it is possible to draw three main conclusions about partnership-based policing in rural areas.

First, questions can be raised about the extent to which the governance of crime prevention has changed. This chapter has revealed some of the mechanisms of governmentality that are used to encourage, coerce or oblige local agencies to take responsibility for local policing. In the UK legislative frameworks form the basis for co-opting local agencies into local policing networks but their effectiveness has been limited according to space and scale. It is possible to conclude that rural places have fallen through this legislation, mainly because it relied too heavily on parish councils to implement it. As wider attempts to reform this organisations have proved, parish councils are too patchy in their effectiveness and fail to represent fully many rural people (Gardner 2008). As Woods and Goodwin (2003) remind us, changes in rural governance cannot be assumed but, instead, rely the micro-politics of participation that can vary significantly within and between rural places. It remains to be seen whether local policing teams introduced by the Police White Paper will also flounder in some rural places as gaps in governance and local democratic participation emerge.

By contrast, the Western Australian example made no use of legislation but, instead, relied on fiscal incentive and programme auditing to ensure not only rural shire councils participated in local crime prevention programmes but they did so in ways that were deemed suitable by state government. Such mechanisms are a clear illustration of 'government from a distance' (Herbert-Cheshire 2000) and, like the UK example, question the extent to which decision making has shifted from government to governance.

Second, the WA and UK examples suggest that little has changed about the responsibility for rural policing. The police are still expected to police rural places and, despite legislation, fiscal coercion and the development of 'active citizen' policies, this is still the dominant discourse in these places. By contrast responsibility for policing in rural New Zealand is more widely shared by police and community despite no or very little formal partnership working. Here, voluntary schemes draw on social capital, a tradition voluntary working and a historic acceptance that the state cannot provide full-time emergency services in rural areas. Although New Zealand is perhaps the most politically de-regulated of the three countries studied in this chapter, the propensity to support voluntary working pre-dates neo-liberal reform and does not reflect a state-centred drive to promote voluntarism that has been witnessed in some countries. Although sustaining this voluntary effort is not without problems, it has led to a sense of shared responsibility amongst many people in many places. It is therefore possible to conclude that historic contingencies as much as government steer contributes to the effectiveness of voluntary action and collective responsibility. The spatial and social effectiveness of rural community-policing would appear to depend strongly upon these traditions and policies to police in partnership merely reflect, rather than affect, propensity to work collectively and communally.

Finally, recent scholarly interest in rural crime and policing has been driven by a sense of academic neglect and a feeling that rural policing is in crisis (Hogg and Carrington 2006). It is worth noting that for many decision makers in the rural places, crime remains something on a non-issue. For these people there is no sense of urgency to act upon it, either communally or in partnership. Hence, many government attempts to spread the responsibility for rural policing or encourage local participation have fallen rather flat. Despite speculation that policing is being shared by more groups (Yarwood 2007b), it seems that in rural places the police, albeit underfunded, under-appreciated and often far away, will continue to take the responsibility for holding the blue line in the countryside.

PART II
Policing the Rural

Chapter 10
Policing Rural Protest

Michael Woods

Introduction

This chapter discusses the challenges posed by new waves of countryside protests for the policing of rural areas. It outlines the strategies adopted by protestors and the police and considers the impact of these on relations between the police and the wider rural community. It draws primarily on research undertaken for an ESRC-funded project on grassroots rural protest and political activity in Britain, including interviews with national leaders of rural groups and local rural activists in Wales, South West England and East Anglia.[1] In this respect, the discussion requires two qualifications. Firstly, the account presented here consequently emphasises the protesters' perspective, although information has also been obtained from police reports and public statements and from off-the-record conversations with police officers. Secondly, the analysis in this chapter is limited to Britain, and may not translate to other countries where the dynamics between police and rural demonstrators will be shaped by national political contexts. Rural protests on the scale of those experienced in Britain are rare in western democracies (with the notable exception of France), and the responses adopted by police have varied. Riot police were deployed against farm protests in Greece in February 2009, and against farm protesters at international summits in Genoa and Röstock, but in other cases the handling of rural protest has been conciliatory. In France, riot police have been used to break up demonstrations by farmers on occasions, especially in Paris, but the policing of other protests have been more hands-off (Naylor 1994). As such, there is likely to be some resonance with the complexities of policing rural protest in Britain discussed in this chapter, but inevitably also some differences.

---

1 'Grassroots Rural Protest and Political Activity', funded by the Economic and Social Research Council (Award RES-000-23-1317), led by Michael Woods and Jon Anderson, with Steven Guilbert and Suzie Watkin. The interviews quoted in this chapter were conducted by Steven Guilbert and Suzie Watkin between January 2006 and December 2007.

Protest out of Place? Policing and the Spatial Order of the Countryside

Protests in the Countryside 1970–1994

The policing of protest has not traditionally been high in the priorities of rural police forces in Britain. Although there were spates of rural protests during the eighteenth and nineteenth centuries (Mingay 1989), 20th century rural Britain was popularly perceived to be an 'apolitical' society, where life proceeded free from protest and conflict (Woods 2005). The 'apolitical countryside' was in fact an expression of a deeply conservative (and Conservative) political hegemony, maintained by a paternalistic power structures (Newby *et al*. 1978; Woods 2005). Agricultural interests were largely represented through farm union involvement in a close-knit policy community with Ministry officials, with occasional small-scale demonstrations by dissident farmers tolerated as pressure releases that protected the general consensus.

Larger protests in rural localities could be represented as displacements of urban politics, transposed because of the coincidental location in rural areas of targets such as military bases and nuclear power stations. Protest events such as anti-nuclear marches and camps largely engaged urban-based participants and espoused radical left-wing political causes that jarred with the predominantly Conservative politics of rural localities (Cresswell 1996).

From the 1970s, the British countryside also saw the emergence of protest activities that more directly challenged the function of rural space and the practice and politics of rural life. First on the scene were hunt saboteurs, anti-hunting activists who had become disillusioned with parliamentary efforts to ban hunting (Bucke and James 1998; Clayton 2004; Thomas 1983). Hunt followers and landowners called for police action to protect their right to enjoy the legal pursuit of hunting. Yet, many saboteurs became adept at knowing the boundaries of legal and illegal activity so that the ability of the police to remove hunt saboteurs was more limited, generating pressure from the rural elite for police powers to be increased.

A second strand within the animal rights movement directed its attention towards agriculture, especially battery farming and the live export of farm animals. Although non-violent direct action was pursed initially, extremist elements within groups such as the Animal Liberation Front and the Animal Rights Militia embraced criminal tactics. But the most extreme actions were directed at scientific and industrial targets, intensive farms (including fur farms) were also attacked.

Wider support was mobilised for protests against the live export of livestock, especially veal calves. Between 1994 and 1996, daily protests at ports involved thousands of participants and lead to several port authorities attempting to halt live animal exports, threatening the livelihood of livestock farmers. The powers of the police once again fell short of the expectations of rural elites in the protection of traditional rural livelihoods.

The third arena of protests to emerge from the 1970s was that of environmental direct action, notably in opposition to the construction of new roads. In early cases opposition was led by local middle-class campaigners whose modest demonstrations operated within the law and were supplementary to the main thrust of political lobbying (Bryant 1996). By the 1990s, these campaigns attracted a new cohort of activists, committed to direct action including trespass, criminal damage and protest camps to obstruct physically construction work (Wall 1999). Police were responsible for enforcing the eviction of protesters from the camps in accordance with existing public order and trespass legislation. Once again, elite groups became frustrated at the apparent limitations of police powers to counter the protests.

In all three cases the protests were perceived as primarily by urban-based participants whose transgression, politics and lifestyles threatened established rural interests (Sibley 1997). In short, a consensus built amongst rural elites and the Conservative government that such protests were 'out of place' and that the police needed to be empowered to contain and remove protesters from rural space.

The Criminal Justice and Public Order Act 1994

The vehicle for expanding these police powers was the Criminal Justice and Public Order Act 1994 (CJPOA). The public order clauses of the Act modified the spatial regulation of the countryside in two ways:

> First, it identified 'those who do not belong' and proposed means for their removal. Secondly, it identified movements and migrations which needed to be curtailed. Some of the discrepant or non-belonging groups are also recognised as 'undesirable' migrants, but others, like many environmental protesters, are criminalised only because they move into the countryside, not because they are social outcasts in any other sense. (Sibley 1997: 222)

The Act hence addressed both perceived political and cultural transgression of rural space. One provision gave the police the power to direct trespassers illegally residing on a piece of land to leave the site. This provision was targeted at New Age Travellers (see also chapters by Halfacree and James, this volume) but the measure also impacted on the policing of protest. Another provision of the Act was targeted directly at protesters, introducing a new offence of aggravated trespass. As defined in Section 68 of the Act, this involved trespassing, or the intent to trespass, by two or more people on land in the open air 'with the common purpose of intimidating persons so as to deter them from engaging in a lawful activity or of obstructing or disrupting a lawful activity' (Bucke and James 1998: 34).

The first major test of the new provisions was provided in July 1995 when protest camps were established against the proposed A34 Newbury by-pass in Berkshire. The eviction of protesters commenced in January 1996, representing a shift in police strategy (Merrick 1996). The role of the police was conditioned by the Highways Agency that sought eviction orders through civil law. The responsibility for serving these eviction orders fell to the Under Sheriff of Berkshire who carried out the task with the aid of bailiffs and contracted security guards. Senior police officers initially saw their role as maintaining the peace but their position inevitably changed as resistance to eviction by protesters raised public order issues (Bucke and James 1998; Merrick 1996).

The police were, however, responsible for enforcing the new law on aggravated trespass. Whereas eviction orders were employed to clear established camps (which had some legal protection), the 356 arrests for aggravated trespass usually occurred 'when protesters breached a cordon of security guards and climbed trees or chained themselves to machinery' (Bucke and James 1998: 49).

The application of these measures showed the spatial strategy of protest policing in operation at Newbury as the largest group of protesters were essentially arrested for *being in the wrong place*. This objective of removing people deemed to be 'out of place', was reinforced by the conditions on bail placed on individuals charged with aggravated trespass:

> if you gave an on-route camp as your address, you were bailed not to come within 1km of the rest of the route, and to sign on at Newbury police station every day. If you gave your home address, you were bailed not to come within 1km of the route, to sleep at your home address every night, and to sign on at your local police station until your hearing. (Merrick 1996: 43)

The employment of a spatial strategy at Newbury was assisted by the geographically fixed nature of the route, which permitted finite exclusion zones to be identified. The greater fluidity of hunts, by contrast, presented greater challenges.

By the mid-1990s, hunt saboteurs were attending most hunt meets, but confrontations requiring police intervention would only occur in a minority of cases. As the eruption of disorder could not be easily predicted, police forces needed to make strategic decisions about what resources to allocate to policing hunts, which in turn influenced the ability of police to take action when offences were allegedly committed (Bucke and James 1998). Confusion also arose about the definition of the 'piece of land' that saboteurs had been directed to leave and the status of protesters on public roads and footpaths.

Consequently, the CJPOA had produced increased expectations of policing among hunt followers that could not always be met, leading to criticism of the police from hunts (Bucke and James 1998). Furthermore, police attention was not exclusively directed at the saboteurs. Many hunts had responded to concerns about limited police resources by employing their own stewards, who could under common law ask trespassing saboteurs to leave private land and use reasonable force to eject them. However, 'police officers' views on stewards were usually negative, since their use was viewed as leading to an escalation of the violence and disorder surrounding the hunt' (Bucke and James 1998: 36; Stokes 1996). Collectively, these factors strained relations between police and hunt supporters in several areas, with police accused of not doing their job and protecting the rights of hunters, which in turned coloured perceptions when hunt supporters themselves became the protesters.

Policing Protest from Within: Rural Protests 1997–2005

Farmers' Protests

In the winter of 1997–8, farmers blocked ports to protest against the export ban on British beef imposed at the time of the BSE scare. These came to an abrupt end after violent clashes with police at Holyhead on 28th January 1998. Many of the farmers who had been involved in the protests had been shocked by the turn of events, which had offended their self-image as law-abiding citizens engaged in legitimate protest. The decision to stand down the protest was hence an attempt to calm the situation and to retain the moral high ground.

However, the mood of anger and discontent within the farming community had not been spent. Economic pressures continued to afflict agriculture, with different sectors hit sequentially by falling prices. As dissatisfaction grew with the perceived inability of farmers' unions to protect farmers' interests, more and more farmers followed the example of the Holyhead pickets and embarked on direct action to make their case.

Between the middle of 1996 and March 2000, over 100 protest events had been organised by farmers in Britain (Reed 2004; Woods 2005). These included demonstrations and rallies in town centres, pickets outside supermarket stores and distribution centres, fast food restaurants and food processing plants, as well as symbolic acts such as dumping surplus milk on fields (Woods 2005). In most of these cases the protests were orderly and peaceful and demands on police were restricted to traffic management and ensuring public safety. Some protests, however, began to push the boundaries of legality, including 'rolling road blocks' of slow-moving tractors in South Wales in February 2000 and a blockade of the Second Severn Crossing in September 1999.

The police were generally notified of protest events in advance, and organisers frequently followed police advice on their plans. Yet, the absence of a clear leadership structure for the farm protest movement could hinder communication with the police. Initially, protests were organised by collective grassroots decision-making within autonomous local groups that were loosely connected. Protests spread around the country by imitation rather than by central direction. Even with the formation of Farmers for Action (FFA) in May 2000 organisation remained largely decentred. The amorphous nature of the farm protest groups exposed their inexperience, but could also help to buy time for protest activity in negotiations with police:

> We were very clever; we boxed very clever. We played the line of the law as far as we dared. We blocked the motorway; we drove around the roundabout and

all parked up. The first time by accident, after that we learned how to do it. And we knew that we could block a highway for 20 minutes before the law could do anything about it. So we played that line. Then the Police Inspector would turn up. The traffic police realised there was a problem then they'd have to go and get the duty Inspector out. So the duty Inspector would turn up and he would say, 'Well, who's the leader? 'Haven't got one' … 'Who's your spokesperson?' 'Haven't got one'. 'Well, you've got to have a spokesperson, 'cos we need to speak to you'. 'I'll tell you what …' and we used to play around whoever was there, 'Well, we'll go over to the other side of the motorway or dual carriageway and we'll have a meeting, and we'll nominate a spokesperson'. Remember, you've still blocked everything up at this point in time. You're talking, you are not being anti. So the Inspector was happy, reporting back that we're talking. (Farm protest activist, Devon, interview)

A further challenge for police came from the spontaneity of some protests. Although many events were organised in advance, some of the most disruptive were spur-of-the-moment decisions by heated meetings of farmers, which caught the police off-guard. North Wales Police, for example, attributed the failure to make any arrests following the ransacking of the Irish lorry at Holyhead on the first night of the 1997–8 blockades to their unpreparedness. In at least some cases the police were warned in advance:

So at 10 o'clock on a Sunday morning, after we'd realised what we were going to do, I took the choice … to phone all the media in the South West and several other big press organisations to say, right, as from eight o'clock on Monday night we will be closing the fuel depots at Plymouth. I also, to be honest, phoned Special Branch and told them what I was going to do, and they seemed a little surprised, but they were waiting. And I was asked, 'What do you want us to do?' And we said, 'We just want a few police there. That will keep it minimal, that side of it.' (Farm protest activist, Devon, interview)

Notifying the police in advance shifted the onus on to police to decide how they would seek to engage protesters. In the case of Plymouth, this led to the police closing off the fuel depot in anticipation of the proposed blockade:

At six o'clock I got a phone call from three or four of the lads to say … 'I thought you said that you wasn't coming down here 'til eight o'clock?' I said, 'why?' They said, 'It's closed'. I said, well what do you mean? They said the police put barricades right the way round all the gates of all the depots and shut 'em. I said pardon? They said, 'yeah, the police have shut the whole lot'. Fine. So I phoned up my contacts in the police – yeah, they're all shut. Don't worry about it. We turned up now and the police were there. I asked to see the duty Inspector and he said, 'What do you want to do?', and for miles, queues of vehicles wanted to come and protest, right down to people in invalid carriages

– that's what shook me. So I said, 'Obviously, we'll have a big drive round for the cameras for the nine o'clock [news], another one at 10 o'clock. I said, you know, we don't want all these police here, do we, all night. If you're happy to keep half a dozen policemen here, we'll keep just a handful of us here, send the rest home and tell them to come back on shifts.' Well no problems at all. (Farm protest activist, Devon, interview)

The implied cooperation between police and protesters in negotiating an 'acceptable' level of protest at Plymouth was repeated at other sites and aimed to respect the right to peaceful protest. Yet, as the impact of the fuel depot pickets combined with panic-buying by consumers led to an estimated 90% of petrol stations running out of fuel within six days, and as the situation escalated into an economic and political crisis, the policing of the protests became of contention.

Government ministers, Members of Parliament, and trades union leaders all accused the police of failing to take appropriate action to remove protesters and to prevent intimidation of tanker drivers, with several contrasting the policing of the fuel blockades with the policing of the miners' strike in 1984 (*Daily Mail*, 15 September 2000; *The Guardian*, 13 September 2000; *The Mirror*, 18 September 2000; *The Times*, 15 September, 2000). Police leaders responded by pointing out that the legislation facilitating police action in industrial disputes such as the miners' strike did not extend to the fuel protests, and reminded politicians that local police forces were 'operationally independent' and could not be directed by central government (Doherty *et al.* 2003; Hopkins 2000). Equally, the fuel companies did not initiate any action against the protesters, such that the police were not called on to help enforce eviction orders.

The offence of aggravated trespass was not invoked during the fuel protests, which reflected the stated position of senior police officers that they did not wish to get drawn into actions that could be perceived as politically motivated (Hopkins 2000). Instead, the police emphasised a commitment to 'policing by consent' and pointed to the absence of violent confrontation at the protests as evidence that a strategy of 'managing' the protests had been successful.

The contrast of this approach with the more confrontational tactics employed with environmental direct action protesters a few years earlier is striking, but suggested differences both in the protesters' self-image and in their perceptions of the police. Drawn predominantly from Conservative backgrounds, the farm protesters lacked ideological antagonism towards the police and often appeared embarrassed to be in a position where they were facing them. Protest leaders have claimed that they insisted on lawful action and enforced discipline from within:

I said, 'you remember you are all business people. You are not thugs, you are not gangsters, you are not hooligans. You are going there to draw attention to the government of the day that you have a legitimate cause. Now if you go untidy and violent', I said, 'you will get no sympathy at all'. I said, 'Let's try it. Now, providing you're prepared to go down that avenue, because I'm telling you now, if you agree, the first one that steps out of line, I'll deal with him first and then hand him over to the police. (Farm protest leader, North Wales, interview)

Similarly, the consensual approach reflected the police's perception of the farm protests. Crucially, whilst the protests were understood as potentially disruptive actions that needed to be managed and contained in space, they were not considered to be out of place, either in the countryside or in the liminal spaces of ports and fuel depots. The farmers' right to protest about the condition of a central facet of rural life was accepted and respected.

Hunting Protests

Even when compared to the farm activists in the previous section, hunting supporters were novices at protest. A survey that we conducted of Countryside Alliance members in four regions found that only 10% had participated in a protest before 1997, but that 75% had joined the Liberty and Livelihood March organised by the Countryside Alliance, and 17% had engaged in direct action in defence of hunting or in support of farmers.[2] For members of the hunting community to suddenly find themselves as protesters and as the subject of policing, was consequently a dramatic and sometimes uncomfortable reversal of roles.

The tension between the desire to protest and the instinct to be restrained and law-abiding was evident in many of the mainstream pro-hunting demonstrations. Maintaining this balance was important to the Countryside Alliance's strategy: disciplined organisation was a key factor in enabling the mobilisation of large numbers of people in predominantly peaceful mass demonstrations that received largely positive media coverage, and thus to stake claim to the 'moral high ground' in the struggle with hunting opponents.

In addition to the three large-scale demonstrations in central London of 1997, 1998 and 2002, the Countryside Alliance organised or supported a number of regional rallies and demonstrations, protests at party conferences, beacon lightings and long-distance marches. Many of these events took place in the countryside itself. All were characterised by the meticulous planning, including close cooperation with the police. Accordingly, as with many of the farm protests, the role of the police at the mainstream pro-hunting demonstrations was essentially restricted to traffic management and crowd control, rather than more politicised actions such as the requirement to stop, remove or arrest protesters.

---

2   The survey was conducted with the cooperation of the Countryside Alliance in summer 2007, as part of the ESRC-funded project on 'Grassroots Protest and Political Activity'. Postal questionnaire surveys were sent to members of the Countryside Alliance in four localities: North Devon and West Somerset; Cheshire; Suffolk; and Mid and West Wales. A total of 1,243 useable questionnaires were returned, representing a response rate of approximately 29%.

Measured protests, however, brought measured results. The early Countryside Alliance demonstrations had caused the Government to panic and to pull-back on support for the Foster Bill on hunting. Yet, as political priorities changed in the new millennium eventual Government sponsorship of anti-hunting legislation started to look increasingly likely (Woods 2008). At the same time, frustration mounted in the hunting community at the moderation of the Countryside Alliance strategy, with calls for more radical direct action.

The first major breakaway group to emerge was the Countryside Action Network (CAN), formed by the former Countryside Alliance spokeswoman, Janet George, which claimed to have 4,000 supporters. During 2002, CAN organised a series of protest events which primarily targeted major roads, blockading the Severn Bridge and streets outside Parliament, and mobilising 'rolling roadblocks' of slow-moving vehicles. Lawyers working with CAN produced guidance for protesters engaged in such actions, noting that, 'It's the sort of thing Greenpeace has got and the hunt saboteurs have got, but nobody inside hunting has got' (Walton 2002: 50).

More provocative and higher profile was the Real Countryside Alliance, or Real CA, launched in May 2002. Deliberately aping the title of the Real IRA, the group adopted 'the language of bombers and hijackers' (*The Guardian*, 30 August 2002) in describing a structure of autonomous cells with each activist having limited knowledge of other activists The anonymous spokespeople for the Real CA promised 'aggressive disruption' and civil disobedience (*Daily Telegraph*, 5 May 2002).

The secretive approach of the Real CA gave it a fluidity that led to the group being associated with a range of protests undertaken by militant pro-hunting activists. These frequently involved acts of trespass and minor criminal damage, including defacing MPs' offices and road signs with pro-hunting stickers, daubing graffiti on bridges, and media-focused stunts such as placing papier maché figures of huntsmen on the white horses at Kilburn and Uffington, and suspending a banner reading 'Love Hunting' from the Angel of the North. Significantly, the police were publicly more critical of these actions than they had been of earlier protests such as road-blocks.

An article published in *The Field* magazine warned of greater disruption including sabotaging reservoirs and blocking motorways (Walton 2002). These threats were not realised, but they were taken seriously enough by the police for the National Criminal Intelligence Service to compile a report assessing the risk. As had been observed previously in relation to the fuel protests, the police's ability to anticipate radical rural protests was seriously hindered by a lack of intelligence on rural militants and their intentions.

Moreover, whereas earlier rural demonstrations had been organised in consultation with the police, the new militant protesters were more selective in their decisions to notify the police. One pro-hunting activist in South Wales, for example, boasted of 'catching the police out' in picketing the house of government minister Peter Hain, but admitted the necessity of advising the police about the blockade of the Severn Bridge:

> Well we had to, because if you do it without telling the police it causes hassle. The police didn't really want to know, but in the end they had to. Well, we said if you don't help us we'll go and do it anyway. And they said right, and we met in a field, just off the motorway and went from there … So no one was going to get hurt, it was just going to make a protest. We're a bloody nuisance really … I mean, certain things you would have to tell the police. (Pro-hunting activist, South Wales, interview)

Perhaps as a consequence, arrests linked to militant pro-hunting protests were limited.

The growing militancy of hunt supporters was expressed not only in direct actions within the countryside but also in attitudes to protests in London. These mirrored the previous manifestation of urban-led protests in the countryside by taking rural protests into the city, embodying the idea of a rural-urban divide (Woods 2005). On the one hand the Countryside Alliance explicitly presented the protests as being 'out of place', involving simple rural folk forced to come to the big city to defend their rights. Yet, on the other hand, the Countryside Alliance also discursively positioned the protests in the tradition of previous mass rallies, including anti-nuclear demonstrations and anti-poll tax protests, to suggest that as protests they were happening in precisely the right place. The police similarly did not consider the London rallies to be 'out of place' and so approached the rallies in the same way as other large-scale demonstrations, deploying additional personnel (1,600 extra officers for the Liberty and Livelihood March), and imposing a schema of spatial regulation that channelled participants along the prescribed route.

The London rallies were therefore examples of 'negotiated protests', in which authorities accept and even assist the legitimate right to protest in return for protesters complying with regulation that limits the disruptive impact of such events. However, this accommodation was criticised by some activists and in response, militant groups organised direct actions with shorter notice that subverted spatial orderings and tested police tolerance, including releasing hounds in Parliament Square and laying turf on roads in Westminster (IPCC 2006). The more assertive protest tactics were met by more forceful policing, culminating in violent clashes between demonstrators and police at a Countryside Alliance organised protest outside Parliament on 15 September 2004, as the Hunting Bill received its second reading. These challenged the reputation of the rural demonstrations as being peaceful protests, and also created friction between countryside activists and the police, further complicating an already complex relationship, as the next section discusses.

The Problems of Policing Local Protest in an Unsettled Countryside

The rise of countryside protests by farmers and hunting supporters required a shift in mentality by rural police forces, from conceiving of protests in rural space as resulting from the incursion of urban political issues and urban activists who could be dealt with through spatial containment, to having to respond to protest originating from within the rural community. In addition to presenting significant logistical challenges, the new dynamics of rural protest drew into question the relationship between the police and members of the rural community who were engaged in protest activity.

The nature of this relationship was complex. Firstly, the more limited resources of rural police forces meant that the officers engaged in policing protests were frequently individuals who were also involved in the more routine policing of rural communities. In particular, some officers may have worked closely with hunt officials in policing saboteur activities at hunt meets, whilst others may have worked with farmers through schemes such as Farmwatch. Secondly, the rural protests were notable for their engagement of establishment figures and included magistrates, councillors and members of police authorities. Some will have had an involvement in the operation of the policing and legal system, whilst others will have moved in elite networks that brought them into contact with senior police officers in their locality (Woods 2005). Thirdly, rural police officers are also usually rural residents, and may therefore have had some sympathy with the causes espoused by protesters. Some may have participated in hunting themselves, or had relatives in farming.

The problem of policing local protest, especially in rural areas, has been recognised before. Baker (2007), in a study of industrial disputes in rural Australia, notes that aggressive policing tactics are more likely to deployed where 'outside' forces were brought in to assist policing than where disputes are handled by local police with a stronger connection to the community.

The complex inter-connections between the police and rural communities led to a widespread belief among protesters that the police were really on their side. This conviction was employed by rural protesters to distance themselves from previous confrontational protests by environmental activists and peace activists and to suggest that their protests were peaceful events at which a police presence was a mere formality:

> It wasn't like a miners' strike or anything like that. There was no aggravation between use and the police, laughing and joking … Everyone was on side … The police said they love demonstrations with farmers. There was never any aggravation. If one of our lads got a bit heated we looked after them, sorted it out. We don't want the police involved. (Farm protest activist, Devon, interview)

However, there were also geographical variations in policing, with peripherality seemingly a key factor. In the port blockades of Winter 1997/8, police from the largely rural Dyfed-Powys and North Wales forces established working agreements with demonstrators at Fishguard and Holyhead, permitting small groups of farmers to inspect incoming trucks and advising drivers at Fishguard to turn back. Kent police, in contrast, actively prevented farmers from mounting a blockade at Dover, including making arrests and impounding vehicles.

In line with previous accounts, aggressive policing and confrontations at the protests were associated with the introduction of police from outside the locality, for either political or logistical reasons. The nature of the fuel protests in September 2000 changed with the deployment of specialist forces, as two of the protest leaders recalled:

> The way that the police brought the convoys in, and the way that they marched 200 riot squad down the hill on the lunchtime at Plymouth docks. They flew them in from London. That's when the Home Office took over, and the local police were disgusted. (Fuel protest activist, Devon, interview)

> We got to the Thursday and a senior police officer said, 'It's totally out of my hands now', he said, 'I've had it direct from Westminster', he said, 'from Whitehall'. He said, 'You're to be smashed and discredited' ... And, he said, 'I'm sorry', he said, 'that we've got to put some thousand police officers and dogs and everything here tonight'. I said, 'Listen,' I said, 'we're only here tonight. This is officially the last night. We haven't come for a fight. This is a token gesture.' (Fuel protest activist, North Wales, interview)

The use of outside forces was not always politically motivated, but more commonly a consequence of the limited ability of small rural police forces to devote resources to protest policing over a prolonged period. Rural protesters were also faced with different policing practices when they took themselves outside of their rural localities, most notably with protests in London, including the clashes in Parliament Square in September 2004. The tactics adopted by police on that occasion were no different to those employed for similar demonstrations, yet the resulting confusion and large number of complaints in part reflected the inexperience of the pro-hunting protesters and their sudden out-of-placeness.

Wherever confrontations have occurred between police and rural protesters, they have contributed to the straining of relations between police and rural communities, compounded by wider concerns such as the handling of the Tony Martin incident in 1999–2000 (Yarwood 2008) and the closure of rural police stations. Furthermore, the failure of the pro-hunting campaign to prevent the passing of the Hunting Act and consequently the prospect of police enforcing the ban on hunting also unsettled the dynamics of police-community relations in rural areas, especially as over 50,000 individuals had pledged to engage in civil disobedience by ignoring the ban, with the tacit support of the Countryside Alliance (Woods 2005). These concerns were expressed most keenly by rural police forces that feared that resources would be stretched and relations with rural communities damaged (Orr-Munro 2005).

In the event, hunt supporters elected to exploit the vagaries of the new legislation by testing the limits of the law through 'legal hunting' rather by overt civil disobedience or continued protest. This has permitted rural police forces to concentrate on the policing of the ban as opposed to the policing of resistance by hunt supporters, although this in itself has produced resource and logistical challenges (ACPO 2005; North Yorkshire Police 2008) and has generated occasional conflict both with anti-hunting campaigners, who have accused the police of not enforcing the legalisation sufficiently, and with hunting supporters who have contested arrests that have been made for illegal hunting (*Western Daily Press*, 20 October 2006; *Western Mail*, 31 July 2007).

Conclusion

The introduction of the hunting ban in February 2005 and improvements in farm incomes have in different ways killed the momentum of rural protests. Yet, the protests of the late 1990s and early 2000s have left a clear legacy for rural politics and society. Most significantly, they have created an environment in which protest activity is not seen as intrinsically 'out of place' in the countryside, but rather is accepted as a legitimate tool for advancing rural interests. Local-scale protests continue to develop around issues such as windfarms, new supermarkets or the closure of village schools or post offices, some of which involve veterans of the hunting and farming protests, some of which have been inspired by these earlier protests. As small-scale, largely peaceful events current rural protests pose few difficulties for police, yet there is also a more permissive attitude towards militant action when deemed necessary. Of the Countryside Alliance members that we surveyed in 2007, 50% thought that breaking the law could be justified for the right cause, compared with only 38% who considered the impact of direct action to be more negative than positive.

The police too have changed their perceptions, recognising that protests can emerge from within the countryside as well as from outside, and refining their strategies as a consequence. Even small and remote rural police forces are now better prepared to confront protesters, including measures such as the training of specialist squads able to drive tractors and heavy goods vehicles as a tool for dismantling road-blocks. Friendly accommodations such as those reached in the early days of farm protests and hunting protests are unlikely to be repeated and the nature of rural policing has been irrevocably changed.

Chapter 11

Still 'Out of Place in the Country'?
Travellers and the Post-Productivist Rural

Keith Halfacree

Introduction: Three Snapshots from Southern England, September 2008

'Widespread opposition to an £844,000 short-term travellers' site in Fakenham were voiced yesterday as planners held a meeting on the proposed site', reports the *Norfolk Eastern Daily Press* (2008). The story goes on to emphasise the controversial character of the proposal for a modest site containing ten hard standing pitches in a field next to a main road on the edge of the town of Fakenham. It notes a poll of local residents that has come out with an 'overwhelming objection to the proposals', and concerns centring on lack of detailed information, highway safety issues, countryside encroachment and, crucially, the 'failure to accept a democratic view'.

BBC News (Bone 2008) constructs a longer story entitled 'We have to fortify our fields', which speaks of a rise in crime and anti-social behaviour on farms across England. Farmers and their representatives recount thefts of fuel, fly-tipping (including burned-out cars and 'full bathroom suites'), animals dumped on fields for free grazing, illegal encampments and intimidation. This is framed through blaming the travelling community, although this link is only explicitly made with respect to encampments. Instead, it is done through the story title (a quote about preventing illegal encampments) and an early boxed (thus highlighted). reproduction of a quote near the story's end: 'We've [the police] got to act almost like a buffer between the settled and traveller communities'.

Later in the month, the *Northants Evening Telegraph* (2008) flags up a proposal by planners for East Northamptonshire council to facilitate the creation of an 'environmentally-friendly village' in the Rockingham Forest area so as, according to the Housing Strategy Manager, 'to meet the identified accommodation needs for New Travellers … [who] prefer rural areas with access to countryside and open space'. The proposal, if enacted, would cater for around 12 adults and 20 children and would hopefully 'increase harmony between residents and travellers and promote understanding between the different cultures'.

These three short media stories suggest something of the place of travellers within the British countryside today. This place has much in common with that in the

1990s but now encompasses a sense of greater ambiguity, away from a strong position as rural 'folk devils' to one of sometimes being accepted as one of the diverse groups living in the rural. Nonetheless, the chapter will not argue that travellers are now 'in place'. Indeed, a central argument is that travellers' rural (non-) placing is always ambiguous, albeit with one representational expression typically predominating.

Reflecting this heightened uncertainty, the first example tells an all too familiar tale of resistance to travellers' sites (Bancroft 2000), officially expressed through 'technical' concerns about traffic safety but with the unseen but surely smelt elephant in the room being a degree of anti-traveller prejudice contained within the unpacked 'democratic view' (Vanderbeck 2003). The second example initially appears similar, with the real crimes experienced by many farmers elided and lazily associated with the travelling community: 'rural crime as Traveller crime' (Vanderbeck 2003: 369). Nonetheless, the weakness of evidence presented to support this association seems so clear as to undermine the story, reinforced by the police representative noting that many complaints about travellers are motivated by 'intolerance and a lack of understanding'. Finally, the third story may reflect an especially enlightened attitude on the part of East Northamptonshire council officers but it also suggests a much better understanding of the characteristics, motivations and needs of New Travellers, with the scheme, for example, deliberately attuned to cater for 'lifestyle and woodland work skills'.

The aim of this chapter, therefore, is to (re)consider the (lack of) place of travellers within the contemporary British countryside, a status in part reflected in contemporary legislation. This was a topic I considered 13 years ago but this reconsideration is undertaken within a specifically historical-geographical framework that was previously lacking. By this point, however, the reader may be wondering what these issues have to do directly with the present book's concern with policing, rurality and governance. This connection is outlined next.

Policing Travellers[1]

When considering the policing of travellers, a key question from the outset is why they supposedly require specific attention (Hester 1999). Although travellers have over a longstanding period been contentiously linked with specific crimes, notably various kinds of theft (Holloway 2005; Mayall 1988), of equally enduring attention is their supposed threat to public order. For example, analysis of parliamentary debates surrounding the introduction of the Criminal Justice and Public Order Act 1994 showed clearly how travellers (real or imagined) were routinely

---

1   This chapter adopts a loose idea of 'travellers', which should be seen to incorporate as many or as few of the numerous candidate groups as is seen relevant to the precise point under discussion. As Turner (2000: 72) put it, with regard to defining a Gypsy, 'Probably, in the end, the most valid indicator is self-definition'. See also James in this volume.

constructed as a key public order problem (Halfacree 1996a, 1996b; Sibley 1981). They were presented, using the celebrated terminology of Stanley Cohen (1972), as one of society's 'folk devils', a group causing a 'moral panic' within polite (*sic.*) society. This status can be seen as heavily rooted in the longstanding and globally widespread prejudice against travellers (Bancroft 2005; Hawes and Perez 1995; Lowe and Shaw 1993; Mayall 2004) but its virulent re-expression in the 1980s and 1990s was strongly stimulated by the emergence of what were soon labelled by the press – a key stage in the production of folk devils – 'New Age Travellers' (Hetherington 2000; McKay 1996). Coming out of the free festival and squatting scenes and from largely rural anti-nuclear weapons encampments, but also crucially driven out of the cities as a result of negative experiences such as homelessness and unemployment (Martin 2002), these New Travellers were soon regarded as one of the socially and culturally deviant groups that the strong state of the New Right was committed to eliminate (Gamble 1994). Their non-Gypsy ethnicity offered no protection by race relations legislation and an attraction to nomadic lifestyles was the icing on the intolerance cake, as will be developed further in the next section.

Having recognised the public order threat that travellers are deemed to pose, the next step is to consider how they are policed. Here, 'policing' is understood not just as a function of the police but as 'carried out by a number of different processes and institutional arrangements' (Reiner 2000: 2), since it is 'more than simply preventing crime and implementing the law ... but refers to the enforcement of codes, standards and ideals held by society' (Yarwood 2007: 2). There are at least three interlinked dimensions relevant here. First, there is the construction and then upholding of specific public order legislation. Thus, for example, both the Public Order Act 1986 and the Criminal Justice and Public Order Act 1994 contained sections specifically targeting travellers (see Burke and James 1994; Halfacree 1996a; James 2004, 2006, 2007a, this volume). Second, there are planning regulations, which can be used to prevent the establishment of traveller sites and/or resist attempts by (former) travellers to settle more permanently (Bancroft 2000; Hawes and Perez 1995). Indeed, these regulations have become more prominent with the Criminal Justice and Public Order Act 1994's removal of the requirement of local authorities to provide sites for travellers in their areas (Sibley 2003). Third, policing also comes from within the rural (and urban) community itself, most clearly expressed, through attitudes, in terms of wariness and suspicion of, prejudice against, and even verbal and physical abuse of travellers (Friends, Families and Travellers 2009; Holloway 2005).

Central to all three of these aspects of policing travellers – including the police themselves – is a strong cultural dimension (Girling *et al.* 2000). If crime is seen as 'ideological censure – applied by one social group to another' (McMullan 1987: 268), then 'public order policing is *not* the maintenance of order, but the maintenance of *a particular order*' (Waddington 2000: 159). Thus, which 'order' is being maintained – an irreducibly cultural and political issue – is absolutely paramount. Further noting how 'criminality, cultural threat and rurality' are all 'socially defined, negotiated and contested' (Yarwood 2001: 202; Yarwood and Gardner 2000), attention must be given, first, to the cultural construction of the place of travellers in British society and then, second, to the more specific issue of whether travellers are predominantly regarded as culturally in place or out of place within particular geographical contexts. To address these issues, the next section of the chapter revisits the central argument of my 1996 papers (Halfacree 1996a, 1996b), written in the context of new legislation passing through the British parliament at the time, before subsequent related academic work on the place of travellers is reviewed briefly.

'Out of Place in the Country' Revisited

'Out of Place in the Country'

The passing into law in 1994 of a new Criminal Justice and Public Order Act was a timely moment to consider the place of different categories of 'travellers' within dominant cultural representations of the normative social composition of the British countryside. This was because the legislation or, specifically, certain parts of its Public Order sections, were targeted at their control or even elimination (Campbell 1995; Hawes and Perez 1995; Lloyd 1993; Woods, this volume). Indeed, subsequent research vindicated fears then expressed by the travelling community regarding the likely impact of this legislation (Bancroft 2000; James 2004, 2006, this volume). In short, it heightened 'conflict between nomadism and sedentarism' (James 2006: 483; also Bucke and James 1998; Niner 2004a).

Drawing on Deleuze and Guattari's (1987) discussion of the nomadic threat to settled or sedentary society, I argued that this threat was fundamentally a geographical issue (Halfacree 1996a). In short, nomads expressed a 'smooth' spatiality, whereby 'points ... are strictly subordinated to the paths they determine' (Deleuze and Guattari 1987: 380), whereas sedentary society is committed to a 'striated' spatiality, sharply bounded and delineated. With settlement, distinctive kinds of human identification emerged, initially two forms (de Swaan 1995) – proximity and family – that reinforced sedentarism. Subsequently, striation of space became especially strongly marked and significant under capitalism, through its intimate association with private property and the profit driven system. The nomad threatens the whole system through his or her potential effacement of relatively fixed delineations and, therefore, must be 'forced back into the system by constraint and violence' (Lefebvre 1991: 382).

The precise way that (rural) space is striated further determines the precise attitude of the state towards different groups of travellers. To appreciate this better, 'rural space' can be seen as having three facets (Halfacree 2006a): 'rural localities' inscribed through relatively distinctive spatial practices linked to either production or consumption activities; 'formal representations of the rural', which refer to the way the rural is framed within the capitalist system; and 'everyday lives of (or in) the rural', which are inevitably diverse. Nomads can be seen as threatening every aspect of this spatiality. They can disrupt the predominant spatial practices, especially through 'disrespect' for private property; they can challenge the everyday lives of people in rural areas, showing an alternative way of living; and they can challenge the predominant ways in which the rural is imagined. Developing the latter, the 1996 papers argued strongly that travellers challenged the 'rural idyll' (Halfacree 2003), a powerful imaginative geography which, *inter alia*, posits a rural 'physically consisting of small villages joined by narrow lanes and nestling amongst a patchwork of small fields ... [and s]ocially ... [comprising] a tranquil landscape of timeless stability and community, where people know not just their next door neighbours but everyone else in the village' (Halfacree and Boyle 1998: 9–10). While ethnically distinct 'real' Gypsies (Sibley 1981), whose nomadism is predominantly for economic reasons, may have a grudgingly recognised, blurred, distanced place on the edge of the idyllic imaginary, most of today's travellers, and especially 'New Age' Travellers, neither ethnically distinct nor largely economic migrants, are firmly 'out of place in the country'.

Subsequent Studies

A small but important body of work has sought since 1996 to consider further the conceptual place of travellers within society (Powell 2008). An 'Othering' perspective has reinforced the perceived negative cultural differences that all kinds of travellers represent. An excellent example is Holloway's (2003, 2004, 2005) research on representations of travellers attending Appleby New Fair in northern England, a major event on the traveller calendar not least due to its role as a horse market. Holloway explores how travellers' Otherness was and is constructed and reconstructed, especially through racialisation and deployment of the 'real' Gypsy myth (Sibley 1981), resulting in them being seen in or out of place in different contexts.

   Othering can be traced to Foucauldian ideas of the gaze and social control. One line of this argument is the functionalist suggestion that the settled 'need' travellers to reinforce and reproduce their own norms and fears (ni Shuinéar 1997). Hetherington (2000: 19) proposed the traveller-as-stranger as a blank 'underdetermined figure', always partially known but critical in reinforcing cultural symbolism and boundary making. This is because 'by introducing something that is not expected, a sense of order is established' (Hetherington 2000: 21). There are consequences for travellers themselves in fulfilling this role. Richardson (2006; also Turner 2000) developed this in her exploration of how discourses implicating travellers are highly selective, derogatory (see also Gilling, this volume) and feed into negative policy recommendations on the need for sites, for example. Echoing much found in debates surrounding the 1994 Act (Halfacree 1996a) and earlier (Sibley 1981), key themes in parliamentary debate remain those of 'mess', 'cost' and the supposed 'unfairness' to the settled population of pro-traveller (*sic*) attitudes in planning (*cf.* Niner 2004b).

A recent sociological contribution by Powell (2008) asserts that issues of unequal power relations and the associated (dis)identification of groups, both of which lead to stigmatisation, merit greater attention. Observing how even travellers who settle are still vilified (ni Shuinéar 1997; Vanderbeck 2003), he argues that 'nomadism is not, on its own, a sufficient explanation for [their] continued vilification' (Powell 2008: 91). Neither is the equally well noted ethnic prejudice sufficient. Using an analytical framework inspired by Norbert Elias's work on established-outsider relations, he demonstrates how 'identification *of* Gypsies, disidentification *from* Gypsies and avoidance of all identification *with* Gypsies are the necessary conditions for the maintenance of power differentials ... and the resultant stigmatisation' (Powell 2008: 97).

Finally, a number of studies have shown something of how both the rural population and the police can regard travellers not just as Others but as an actual or potential public order problem. For example, in her investigation of 'white rural residents' discussing 'Gypsy-Travellers', Holloway (2005) shows how a predominant racialisation lens, indicated through both bodily and cultural markers, posits an elusive, heavily idealised 'true Gypsy' (see also Sibley 1981) against the highly resented 'hangers on', the latter associated with criminality and general public order problems. Elsewhere, a study in Worcestershire (Yarwood and Gardner 2000) found the rural population split evenly between those who saw 'travellers' as a crime problem and those who did not. Complementing such studies are some explorations of police attitudes towards travellers. These demonstrate that whilst there are some, perhaps many, officers who show a good understanding of and even sympathy for travellers (James 2006),[2] many others express considerable levels of prejudice (James 2007a), particularly linked to their perceived criminality (theft, damage, drug use) and public order challenge (Hester 1999).

Overall, what post-1996 work on the place of travellers has in common is a strong sense of place or context (Girling *et al.* 2000). Following Little (1999), Otherness has always to be (re)constructed; the key to the blank figure or 'joker' (Hetherington 2000) is how and where it is used; and Powell (2008: 104) notes how stigmatisation is 'reproduced through the spatial order'. Thus, whilst all of these studies regard travellers as mostly negatively Othered, its extent and the form it takes is not fixed and unchanging. This can be developed into a critique of the 1996 articles for, although expressing a degree of contextual sensitivity, this was not developed sufficiently. In part, this reflected a concentration on the representational dimension of rural space. Consequently, the historical-geographical dimension of travellers' place in the country merits revisiting. This is done in the rest of this chapter within the specific setting of the changing spatial order of the British countryside.

---

2   Here it can be noted that the police were not the main driving force behind the Criminal Justice and Public Order Act 1994, for example.

The Pre-productivist Countryside: No Place for Travellers?

This section briefly considers the place of travellers in what will be labelled, very reservedly, the 'pre-productivist countryside' prior to the Second World War and the subsequent drive for a productivist rural. Clearly, such bracketing cannot do justice to 450 years of experiences but it places the next two sections in some historical context.

In considering travellers in the pre-productivist countryside, the widely known and detailed persecution they faced over this whole period (Bancroft 2005; Hawes and Perez 1995; Mayall 1988) suggests strongly that they were accorded little legitimacy within rural space. Although Gypsies were persecuted as 'foreigners' (Okely 1983: 4) almost as soon as they arrived in Britain in the early 16th century, this soon switched to an emphasis on their perceived 'vagrancy' as they were increasingly native born (Hawes and Perez 1995). Indeed, extending far beyond any association with Gypsies, vagrancy was regarded by Tudor and Stuart authorities as 'one of the most pressing problems of the age' (Beier 1985: xix). This stemmed, first, from the collapse of feudalism; where people had been strongly tied to a specific place, now a rise in numbers of 'masterless men' (*sic*) was possible. Second, the new social order had 'to be monitored and enforced' (Bancroft 2000; McMullan 1987: 269) if it was to become institutionalised and normalised; a nomadic disposition challenged this. Moreover, with reference to Gypsies, Okely (1983: 53) notes how their 'history is also the history of their refusal to be proletarianised'. Third, vagrancy was largely a consequence of 'profound social dislocations – a huge and growing poverty problem, disastrous economic and demographic shifts and massive migration' (Beier 1985: 3). The poor were becoming more numerous and troublesome (Sharpe 1988).

As a result of these concerns, numerous pieces of legislation were enacted to deal with vagrants and travellers, with a desire to 'civilize the homeless vagrant' (Hester 1999: 131) a key stimulus to the development of modern policing. Elizabethan law had defined wanderers as 'rogues' (Cresswell 1996) and this was extended and embellished. Tudor governments passed statutes against Gypsies and until around 1600 these laws spawned numerous punitive measures, from licensing begging to imprisonment to corporal punishment (Beier 1985). The perceived threats of criminality, disease carrying and, crucially, challenges to the new social order saw 'England's rural elites ... terrified of vagrants' (Beier 1985: 48). Clearly, fear of the stranger loomed large and probably helped consolidate the emerging proto-capitalist society.

From around 1600, the penal element within vagrancy legislation declined in importance (Beier 1985), due especially to England becoming more prosperous and stable, with strong economic growth, on the one hand, and the overall success of the Tudor and Stuart measures, on the other. Such a broad trajectory was maintained through the next few centuries (Crowther 1992), notwithstanding continued official opprobrium of vagrancy, reflected in measures such as the Vagrancy Act of 1824, which 'made it an offence to be in the open or under a tent, coach, or wagon without any visible means of subsistence and unable to give a good account of oneself' (Cresswell 1996: 84). Legislation continued to treat Gypsies as rogues and vagabonds, with late 19th century charity and legislation seeking to transform their lifestyles and, in particular, stop their nomadism (Holloway 2003; Okely 1983), which clearly retained transgressive threat.

In short, throughout the pre-productivist period, sedentarism and its emphasis on exclusionary private ownership remained the norm (Bancroft 2000; James 2006) – also expressed through Enclosure Acts, for example (Mayall 1988). However, as fear was controlled, and charitable pity became commonplace, a highly romanticised envy of the travelling life developed (Crowther 1992; Sibley 1981). In an increasingly urban, industrial and rational society, the strong cultural associations Gypsies had accreted with freedom, nature, desire and quasi-mystical powers increasingly appealed.

Numerous examples of this romanticism can be cited (Mayall 1988, 2004). However, it will be left to Toad from Kenneth Grahame's 1908 novel *The Wind in the Willows* to express something of the romantic appeal of the nomadic life:

> 'There you are!' cried the Toad, straddling and expanding himself. 'There's real life for you, embodied in that little cart. The open road, the dusty highway, the heath, the common, the hedgerows, the rolling downs! Camps, villages, towns, cities! Here to-day, up and off to somewhere else to-morrow! Travel, change, interest, excitement! The whole world before you, and a horizon that's always changing!' (Grahame 1926: 24)

A second reason for travellers having a place within the pre-productivist countryside is more prosaic. Travellers had a range of important rural economic functions. Gypsies were well established as horse-dealers, blacksmiths, tinkers and scrap metal dealers (Beier 1985; Holloway 2003; Okely 1983). Casual agricultural labour was also important, the seasonally related and geographically varied demands that this brought about being a key structuring device for migrations (Okely 1983). In sum, 'Gypsy-travellers fitted, if not entirely comfortably then at least without excessive conflict, into the rural economic and social structure' (Mayall 1988: 182).

Thirdly, and finally, travellers could be said to have a place within pre-productivist rurality because the British countryside was simply not used too intensively. There was enough space for travellers to establish encampments and not come up against dominant sedentary rural practices. In short, places on the margin were relatively plentiful and it was these that travellers passed through, enhancing their position on the edge of settled society.

The Productivist Countryside: Still No Place for Travellers?

The Second World War had many profound implications for how the British countryside was subsequently reshaped (economically, culturally, politically, socially) but of key significance was a distinctive model of what it was for and how it should develop. Following a well-accepted convention (e.g. Ilbery and Bowler 1998), 'productivism' predominated from around 1945 to about the late 1970s. This positioned agriculture as a ruthlessly efficient forward-looking expansionist food production industry in archetypical capitalist mould.

Crucially, productivism assumed a discourse of rural life: there was a 'productivist countryside' (Halfacree 2006a). Underpinning the practices of productivist agriculture that dominated this countryside were representations of the rural that nominated, normalised and nurtured that countryside as first and foremost a food production resource, and lives for the farming community, at least, that emphasised a strong sense of all round security. Within this 'new' and 'modern' countryside, there was little space for travellers for at least four reasons.

First, the predominant power within the productivist countryside was orientated around the farmers and landowners who were to benefit immediately from the refashioned agricultural industry and their links through networks to the agricultural policy community (Murdoch and Ward 1997). A landscape of supposedly independent producers reliant on modern technologies to rationalise their farms to enhance production could at best posit travellers as an historical anachronism and at worst a nuisance with no legitimate place on the land. This was reflected, almost certainly unintentionally, in rural textbooks of the time, where travellers got no mention (e.g. Clout 1972). It was omissions such as this that subsequently fed into the work on 'neglected rural geographies' (Philo 1992).

Second, a key reason why travellers could be regarded as anachronistic was that, as noted, numerous forms of technology centrally underpinned the new agriculture. Consequently, inputs of casual labour from Gypsies at harvest time, for example, would no longer be required as new harvesting machines assumed the task; in Marxian terms, the fruits of dead labour were to replace living labour power. Travellers were to become an inevitable (*sic*) new chapter in the 'drift from the land'. Thus, even just after 1945 popular writers were associating Gypsies with 'a "vanishing" rural England' (Okely 1983: 24).

Thirdly, the success of productivism, at least in early decades, meant that land left out of production in economically depressed times, some of it used for travellers' encampments, was brought into production. The exclusionary consequences of earlier enclosures could be seen as taking a step further. As has been noted in the context of 1970s back-to-the-land experiments (Halfacree 2006b), whilst the drift from the land may have created surplus farms and houses, land tended to be absorbed into growing agricultural enterprises.

Fourthly, productivism was all about change and development, and within this upheaval one might imagine some travellers being able to slip by. However, as Murdoch *et al.* (2003) argued, throughout productivism's currency a contradictory romantic, communitarian and ideological moral stewardship vision of farmers and landowners overlaid its representation. Within this pastoral and highly sedentary vision of a fixed, settled landscape, as shown above, there was little space for travellers. Indeed, the deterritorialising nomadic threat to the uneasy compromise between pastoralism and modernism would be especially undesirable for those attempting to hold the two currents together.

Nonetheless, although travellers seemed to have no place within the productivist ideal, in practice exclusion was nothing like so clear-cut. First, many farmers still needed casual labour, and Gypsies continued to fulfil this role. Second, all waste land was not put under intensive production. Indeed, productivism expressed a very strong geography, associated with its uneven development within different sectors of agriculture. Thus, sites were still found by travellers, albeit with the hostility and resistance that camps almost always engendered. It should be noted that the Caravan Sites Act 1968 (Hawes and Perez 1995; Sibley 1981) was passed during the productivist era, which, in principle at least, sought to establish a network of Gypsy sites across Britain. Thirdly, whatever the material situation of travellers in Britain, their romanticisation continued (Sibley 1981, 1992). This allowed 'real' Gypsies to retain their place on the edge of the rural idyll; the cultural association with the countryside remained even as Gypsies were becoming increasingly urbanised (Sibley 1981).

In summary, and reflecting Powell's (2008) intervention, in spite of the qualifications made in the previous paragraph, the reality of power relations within the productivist countryside meant that travellers were predominantly out of place. The practices of productivism, its spatial imagination, and the lives led within such a landscape had very little room for nomadic alternatives. If they were to exist, they were to be controlled tightly through the Caravan Sites Act 1968 and the subsequent licensing measures of designating 'legitimate' sites (Sibley 1981).

The Post-productivist Countryside: A Place for Travellers?

Although disagreeing substantially over details, most commentators agree that productivist spatiality has a much less hegemonic hold today than in the 1950s. Moreover, this weakened grip is not just the consequence of the devastating agricultural crises that have impacted so heavily on British agriculture over the past 25 years. Indeed, Drummond *et al.* (2000) spoke of productivist British agriculture being in 'structural crisis', unable to be cured by the 'technical fixes' that rescued it from previous 'conjunctural crises'.

All three facets of productivist rurality have been undermined (Halfacree 2006a, 2007). Its representation was unable to achieve total dominion over either localities or everyday lives and rivals have grown more significant. Rural localities have experienced substantial economic restructuring and social recomposition. Farming has been forced to adjust in response to the varied contradictions of productivism, from surpluses and over-production to perceptions of environmental despoliation. Farming lives are increasingly insecure and uncertain, with any 'guardians of the countryside' role disputed.

Expressing these changes, an agricultural 'post-productivist transition' is characterised by three 'bipolar dimensions of change' (Ilbery and Bowler 1998: 70): intensification to extensification, concentration to dispersion, specialisation to diversification. This emphasis on diversity has extended beyond agriculture, prompting talk of 'post-productivist countryside' (Halfacree 2006a, 2007) where 'agriculture exists within and is encompassed by rural space and society rather than the other way around' (Gray 2000: 42). The spatiality of the post-productivist countryside is fundamentally *heterogeneous*. Consequently, a place for travellers looks at first sight plausible, for at least four reasons.

First, the sheer absence of any hegemonic rurality and, moreover, a *leitmotif* of heterogeneity/diversity speak of an opportunity to be seen as part of the emerging rural. Indeed, trends within both academia (e.g. Philo 1992) and the policy community (e.g. Department of the Environment, Transport and the Regions 2000) to recognise and, to some degree, support diversity within the countryside and counter the Othering of groups such as travellers, can be seen to open up at least an advocacy space for travellers. This is reinforced with the growth of supportive and campaigning traveller networks such as Friends, Families and Travellers (2009) and professional advocates such as Greenwich University's Thomas Acton or Professor Luke Clements at Cardiff Law School (James 2007b).

Second, one of the key characteristics of the emerging post-productivist countryside is the increased recognition and enhanced importance of the rural as a consumption space. No longer are (productivist) agriculture and related forms of production the only legitimate use for the rural. Given that, for example, 'perhaps the key [positive] attribute of travelling [for New Travellers] was ... the ability to move on and into the countryside' (Davis 1997: 125), the diverse rural consumption practices of travellers may be seen as part of this broad consumption mosaic.

Third, and developing a more explicitly representational dimension, arts and the media increasingly include travellers, especially New Travellers, within their rural coverage. In particular, no rural drama/crime series seems complete without an episode featuring travellers, albeit often highly stereotyped. Significantly, whilst an initial association between travellers and criminality is typically made, it is usually the case that this was a false or at least problematic association, based more on characters' prejudices than on the facts. Notwithstanding critiques of such representations – and there are many – they are at least 'normalising' a place for travellers within the British countryside.

Fourth, a further aspect of rural change associated with post-productivism is the greening of agricultural production, typically supported by European Union funding. New Travellers, in particular, are often well placed in this respect, with many keen to settle (Niner 2004a) and undertake permaculture and related forms of production (Halfacree 2007); showing an affinity to working with wood (see the *Northants Evening Telegraph* (2008) article discussed above).

However, sensitivity to Powell's (2008) power perspective means that any potential place for travellers in the post-productivist countryside must be qualified strongly. Although heterogeneity may lie at the heart of the post-productivist idea, a central question is *which differences* will comprise (or dominate) this diversity? It is extremely naïve to suggest that all interested groups are likely to make their mark significantly on the British rural of the 21st century (Halfacree 2007). As noted by Murdoch and Pratt (1993: 425), 'Some rural experiences ... work powerfully to subsume others' and thus to construct any '*bricolage*' (Murdoch and Pratt 1997: 55) of 'rurality' from various Others would fail to recognise this 'topography of power' (Murdoch and Pratt 1997). Critically, changes taking place in the reformatting of rural space are driven primarily by profit-driven involution, the capitalist reworking of a space (Katz 1998) already 'colonised' by productivist capitalism.

Looking more closely at the dominant contours of the emerging post-productivism, two spatialities appear especially powerful (Halfacree 2006a). First, there is 'super-productivism' that reaffirms productivism in a more explicit form. A naked capitalist production 'logic' is apparent in the practices of agribusiness, genetic modification of plants and animals, and biotechnology generally. Land is represented solely as a resource for profit maximisation through globalised industrial agriculture, with 'nature' as 'an accumulation strategy' (Katz 1998). Such is the physical impact of super-productivism that everyday lived rurality sharply converges with this representation. Within all three spatial aspects there is no place for travellers, yet it is a powerful spatiality given its orientation to the global marketplace and the continued dependence on (rural) land for food production, and its presence may well be boosted by the ongoing global 'food crisis'.

'Consuming idylls' is the second key emerging element of post-productivist spatiality. Whilst its locality has an agricultural façade, its dominant spatial practices involve consumption: leisure, residence, in-migration, dwelling, contemplation. Its key representation is the rural idyll (Halfacree 2003). However, the extent to which everyday lives fit with this representation is varied. Indeed, representational and material struggles feature strongly in the creation of an idyllised countryside, not least in the planning and policing spheres. Travellers will also feature in these struggles, not least as they provide the key (nomadic) threat to the spatiality of the rural idyll, as already noted, suggesting that overcoming what the *Northants Evening Telegraph* (2008) termed 'cultural misunderstanding' is unlikely to be easy. Indeed, for a spatiality less coherent and robust than super-productivism, the stranger-ordering role of travellers may prove fecund.

Rural space demonstrates lively capitalistic involution re-working but within all this busi-ness (*sic.*) there is little left for travellers. The sense of rapid change can promote increased levels of suspicion, especially towards competing interests and/or already stigmatised groups, such as travellers (Hester 1999). Consequently, as the present book attests, we now have a heavily policed countryside, supported by a raft of legislation from the Criminal Justice and Public Order Act 1994 to miscellaneous other laws (James 2006, 2007) that can be brought to bear on groups considered deviant presences, such as travellers. Notably, this policing exists in the context of social and economic upheavals prompting heightened fears of crime (Yarwood 2005), in turn often associated with travellers (Vanderbeck 2003). Given this context, the following chapter explores the impact of policy and policing on travellers in more detail.

In summary, whilst our era of 'mobilities' (Urry 2007) may well suggest further that travellers have a place in any post-productivist countryside, this remains at a rather abstract and hypothetical level. Significantly, and ominously for travellers, the dominant forces shaping this new countryside suggest strongly that, as with 'radical ruralities' generally (Halfacree 2007), any embedding of their potential place is likely to necessitate considerable political and cultural struggle at all levels. In short, the predominant post-productivist countryside that seems to be emerging still has very little room for nomadic alternatives.

Conclusion: Still Unlicensed Difference?

To understand the policing of travellers within rural Britain, both by the police and by society more broadly, this chapter has argued that attention must be given to whether or not dominant cultural imaginations regard (specific types of) travellers as having a legitimate place within a changing rurality. The picture seems, in summary, decidedly mixed.

A recent report by the Commission for Racial Equality (2006) concluded that Gypsies and Irish Travellers are the most excluded groups in Britain today. Thus, any sense that travellers are attaining more of a 'legitimate' place in the countryside in these supposedly more open post-productivist rural times must be treated *a priori* with considerable scepticism. Ethnic prejudice, rejection of Others, fear of the stranger, anti-nomadism and the assertion of power through stigmatisation remain hugely powerful forces shaping everyday practices, representations and lives in our daily spaces.

This chapter has sought to locate the sense to which travellers were and are in or out of place within the British countryside in a more careful historical-geographical context than has been done to date. The following chapter considers more closely the policy and policing aspects of these inclusions and exclusions. Although it has concluded quite pessimistically as to whether the supposedly inherently heterogeneous post-productivist countryside includes a place for travellers in practice, in principle this spatiality has much more scope to provide such a place than did productivism, with its very narrow spatial logic. It is up to all who do not demean the nomadic life – planners, academics, charitable groups, travellers themselves, even the police – to build strong enough networks and resources so as to acquire the power to translate this potential into a reality. Whatever our motives for this, if history has taught us anything it is that the nomadic 'alternative' is never likely to go away and deserves to be respected and accommodated justly (Niner 2004a). Even unlicensed difference does not have to be an 'enemy'.

Chapter 12

Gypsies and Travellers in the Countryside: Managing a Risky Population

Zoë James

Introduction

The previous chapter argued that Gypsies and Travellers have been a constant presence in the countryside of England and Wales for centuries, but their presence has nearly always engendered conflict with settled communities (see also Kenrick and Clark 1999). Drawing on an empirical study, this chapter goes on to consider the risks experienced by Gypsies and Travellers in the countryside and how Gypsies and Travellers are perceived as risky by authorities in the countryside. These contested perceptions of risk result in conflictual policing that fails to fulfil the needs of either the settled or Gypsy and Traveller communities.

Policing and Policy

In 1968 The Caravan Sites and Control of Development Act required local authorities to provide sites for Gypsies and Travellers to live on following the closure of the commons that had traditionally served as their stopping places. However, local authorities largely failed to fulfil their duty to provide sites and Gypsies and Travellers found themselves increasingly having to stop and stay in places that were illegal. The enactment of the Criminal Justice and Public Order Act 1994 repealed the requirements of the 1968 Act augmenting the problem of illegal stopping places (James 2004). Three broad categories of illegal site can be identified: illegal developments, unauthorised encampments and roadside sites. Illegal developments are sites that have been developed by Gypsies and Travellers on their own land, but do not have planning permission. Research has shown that Gypsies and Travellers are turned down in 90% of initial planning applications (CRE 2006; Williams 1999).

Other illegal sites are classified as 'unauthorised encampments' where they are set up on any land without permission. Such sites can be very different though, as some unauthorised encampments are long term sites, that, though illegal, are tolerated by local authorities following Home Office Circular 18/94 that advised toleration of illegal sites in the absence of sufficient legal sites, while others are short term sites on the roadside, or on waste ground that Gypsies and Travellers have had to move on to while in transit or having been moved on from elsewhere. The short term, unauthorised encampments on the roadside or waste ground are those with the worst conditions of living for Gypsies and Travellers and are occupied by Gypsies and Travellers from all cultures (James and Browning 2008; Niner 2004a, 2004b).

The Office of the Deputy Prime Minister (now Department for Communities and Local Government) recognised the position of Gypsies and Travellers' poor accommodation circumstances following extensive lobbying by Gypsy and Traveller activists, and consequently placed a requirement in the Housing Act (2004) for the accommodation needs of Gypsies and Travellers to be specifically measured as part of the broader assessment of accommodation need across all local authorities in England and Wales (Southern and James 2006). This requirement was accompanied however by increased powers for eviction of Gypsies and Travellers from unauthorised encampments in the Anti-Social Behaviour Act 2003. I have previously described this coupling of legislation around provision and enforcement as providing a hiatus moment for Gypsies and Travellers, whereby the provision of sites that the law requires, and is desperately desired by Gypsies and Travellers themselves, could result in their traditional nomadism being permanently curtailed, as the Anti-social Behaviour Act eviction powers do not come into force until comprehensive provision of sites is made (James 2007a). In the desire to get somewhere safe to stay, Gypsies and Travellers may find that the assimilation aim of authorities is being quietly met as they express a need for permanent sites that let them off the merry-go-round of multiple evictions, but subsequently fails to provide for nomadism.

Gypsies and Travellers' accommodation problems have resulted in their commonly living in very poor, overcrowded conditions and under threat of eviction. Recently a comprehensive review of the literature in this area was carried out by Cemlyn *et al.* (2009) for the Equality and Human Rights Commission (EHRC). The EHRC, in its previous incarnation as the Commission for Racial Equality (CRE), recognised some time ago that Gypsies and Travellers were experiencing a marginalised, excluded lifestyle. The Cemlyn *et al.* (2009) text draws together research and writing about Gypsies and Travellers over recent years and is therefore able to identify the breadth of exclusion they experience. The report summarises that Gypsies and Travellers lack decent accommodation, suffer poor health and are not sufficiently educated. This is found to be a result of having been ignored and discriminated against by social services and public agencies. The report states that their review of existing literature identifies one key issue across all areas of the study: 'the pervasive and corrosive impact of experiencing racism and discrimination throughout an entire lifespan and in employment, social and public contexts' (Cemlyn *et al.* 2009: iii). Importantly, the Cemlyn *et al.* (2009) text includes a section on Gypsy and Traveller experiences of the criminal justice system and by doing so identifies the fact that Gypsies and Travellers are drawn into this system increasingly as a consequence of their experiences of living on and eviction from illegal sites and as victims of racism and harassment. These two issues will be specifically addressed by the analysis later in this chapter.

The Research

The requirement for local authorities to carry out accommodation needs assessments for Gypsies and Travellers (GTAA) has led to a raft of reports across England that provide the most comprehensive picture of Gypsies and Travellers' needs than has been seen before. Despite government guidance on how these reports should have been produced (ODPM 2006) and identification of best practice (Clark and Greenfields 2006), the GTAAs have utilised a range of methodologies and approaches and have subsequently been critiqued for failing to provide comparable measures of need (Niner 2007). Despite this criticism, the GTAAs are a valuable resource as they provide extensive data on Gypsies and Travellers' experiences that are worthy of further analysis to inform discussion and debate in this area. The GTAAs, though not necessarily comparable in terms of accommodation provision requirements (their official aim), hold information on the type of accommodation inhabited by Gypsies and Travellers and where relevant its legal status. They also outline Gypsies and Travellers' previous accommodation, movement and eviction and they commonly provide information on Gypsies and Travellers' health and welfare status and experience of services, both public and private, including policing and security.

The discussion in this chapter is based on analysis of a sample of 19 randomly sampled GTAAs carried out throughout England between 2006 and 2008. Analysis was carried out using a simple categorisation method, informed by the themes that were drawn from the data at initial review. The themes were then explored in relation to existing evidence and theoretical relevance as will be outlined in this chapter. The researchers reviewed the findings of the GTAAs with a view to drawing out policing experiences and the role of other agencies in managing Gypsy and Traveller communities. In order to fully understand and theoretically consider the findings, it was necessary to additionally draw out the lived experience of Gypsies and Travellers as expressed in the GTAAs, particularly in relation to accommodation. Based on this work, this chapter goes on to identify key issues of control and policing in the countryside as drawn out of GTAAs nationally.

Research Findings

The key finding of this research is that Gypsies and Travellers constitute risk that is framed by contested perceptions (Barton and James 2003), leading to conflictual countryside management. Gypsies and Travellers' lives are considered risky by the Travellers themselves and they are considered risky by the agencies who aim to manage and control them. These risks differ however, according to the group perceiving them, though they are borne of common problems and issues.

Although some Gypsies and Travellers live on authorised, legal sites, the GTAAs identify that Gypsies and Travellers rely on a range of unauthorised encampments. The variable nature of the methodology of the GTAAs, as previously mentioned, has meant that it is hard to compare site status across areas, bar the illegal nature of them. Different GTAAs refer to *'unauthorised sites'*, *'unauthorised camping'*, *'roadside sites'*, as well as the formal term, *'unauthorised encampments'*. Some GTAAs note that the variability of terms used to describe sites is a consequence of the lack of clarity of their legal status – whether Gypsies and Travellers had attained permission from landowners to stay on the land or not and ambiguity over who owned the land. Interestingly, *'farmers fields'* were particularly noted in a number of GTAAs as unclassified stopping places that were not counted within numbers of unauthorised encampments, implying perhaps that such land is publically owned in the public and evaluators' (in this case) conscience. Nevertheless, it is clear that Gypsies and Travellers are heavily reliant on illegal stopping places and they consequently are subject to being evicted on a regular basis.

The process by which Gypsies and Travellers can be evicted from land is set out in the Criminal Justice and Public Order Act 1994, Sections 61 and 77, for police and local authorities to use respectively. This legislation requires particular circumstances to have been met and information provided to Gypsies and Travellers prior to formal eviction. However, there is no requirement for records of eviction process to be kept (Clark and Greenfields 2006; Morris and Clements 2000; Niner 2003), resulting in ad hoc and variably applied approaches by different local authorities and police forces. Moving Gypsies and Travellers from unauthorised encampments is dealt with via a range of methods including formal eviction by police or local authorities, often aided by private bailiff companies; warnings of eviction, sometimes referred to as *'threats'* in order to promote Gypsies and Travellers' movement; negotiation of movement between authorities and Gypsies and Travellers; *'voluntary'* movement within the context of an enforcement process. All the GTAAs analysed outlined eviction as a common experience for Gypsies and Travellers and the large majority went on to note the implications of such movement as detrimental to the health and welfare of the families concerned. One GTAA refers to *'families exhausted from constant movement, and in some cases unable to repair vehicles or even prepare food for their children before having to move on yet again'* (NA15). Another refers to families, *'living in constant insecurity with significant impacts reported on their mental health'* (NA5). One study notes the impact of eviction on families' ability to make informed choices about their accommodation.

According to a number of GTAAs, however, authorities are engaging with Gypsies and Travellers beyond eviction action. Following guidance published in 2006 (ODPM 2006) local authorities have been required to consider the welfare needs of Gypsies and Travellers prior to any eviction taking place. Such consideration does not and has not precluded the process of enforcement occurring, but has been brought into the process of enforcement in the majority of reports in the GTAAs analysed. A response typical to many GTAAs is '*an emphasis on rapid action, on negotiation over leaving dates where necessary backed up by court action for possession following welfare assessments*' (NA4).

Although the process outlined in this instance allows for welfare considerations to be made, it is clear that Gypsies and Travellers are negatively affected by constant movement and are thus likely to experience more problems as a result of rapid movement. The CRE suggested that Gypsies and Travellers were experiencing 'enforced nomadism through constant eviction' (CRE 2006: 4). However, the implication from this is that the provision of legal stopping places for Gypsies and Travellers would negate their need for any movement. This would fail to address the nomadic nature of many Gypsies and Travellers' lives and the importance of such to their cultural understanding, as outlined in the CRE report.

The inclusion of welfare considerations within the enforcement process tends to inform eviction, rather than to provide support or provision for the evicted, and is symptomatic of a multi-agency approach to enforcement that is problematic to Gypsies and Travellers (James 2007a). New Association of Chief Police Officer (ACPO) guidance on the management of unauthorised encampments (ACPO 2008) identifies the need for local authorities to engage with residents of Gypsy and Traveller sites that do not need to be immediately evicted by the police. It also recognises the role of the police in engaging positively with Gypsy and Traveller communities to support them as victims as well as suspects. However, the new guidance formalises the process of engaging with Gypsies and Travellers living on unauthorised encampments as part of an enforcement process. Bancroft noted in 2000 that Gypsies and Travellers exist within a discourse of punishment. Indeed, the analysis here would suggest that they continue to do so. Within the discourse of punishment, under the gaze of authorities (Richardson 2005), Gypsies and Travellers actually experience victimisation from a reticent public and the authorities appear to do little to help. Analysis of the GTAAs shows that the large majority of Gypsies and Travellers, no matter what their origin or accommodation status, have experienced racism and harassment from the settled community. Indeed, this has meant that a large proportion of Gypsies and Travellers consider the harassment they have or may experience as part of their decision on where to live and where to move to.

Gypsies and Travellers reported a range of experiences of harassment to the GTAAs such as, shouting, name calling, stone throwing, physical movement of vehicles while occupied and vandalism. Incidents of harassment were identified as occurring on all types of site, at work and within school environments. One GTAA referred to the '*endemic level of hostility towards Gypsies and Travellers*' (NA4) which was considered particularly problematic due to its normalisation by the Gypsy and Traveller communities who are unlikely to report offences against them. A lack of reporting of offending behaviour is not uncommon amongst socially excluded populations (Rowe 2004), and some work has been done by particular police force initiatives on reporting race hate crimes, such as in the

Metropolitan Police Service, in Cambridgeshire and Merseyside police forces. Analysis of the GTAAs shows that Gypsies and Travellers are unwilling to report such crimes as they are, '*rarely dealt with properly by the police*' (NA14). There is a consequent lack of trust and confidence in the police, who are most commonly associated with eviction and enforcement activity. The GTAAs show that Gypsies and Travellers are more positive about engaging with their local community police officer, reflecting studies on public confidence in the police more broadly (Bradford *et al.* 2009; Moley 2008).

The lack of confidence of Gypsies and Travellers in police services is similarly identified in the GTAAs when analysing relations with other public and private services. As part of the framework of control that Gypsies and Travellers experience, multiple agencies are viewed with suspicion by Gypsies and Travellers in a number of GTAAs. The response of Gypsies and Travellers to the harassment and discrimination they have experienced is to hide their identity from authorities, '*not revealing that they were Gypsies and Travellers*' (NA19). This results in Gypsies and Travellers only engaging with services in crisis situations when problems arise or a practical solution is sought, such as via the use of Accident and Emergency hospital departments. There were a number of positive experiences related in the GTAAs regarding dedicated Gypsy and Traveller services however, such as the Traveller Education Service. Where positive accounts were set out in the GTAAs, they tended to focus on individual workers within an organisation, rather than on the organisation as a whole.

Analysis of the GTAAs shows that conflict is central to Gypsies and Travellers' lived experience, both between Gypsy and Traveller communities and between Gypsies and Travellers and the settled community as initially identified above. It is commonly understood by organisations providing support to Gypsies and Travellers that there is tension between communities and this has been recognised by previous research (James 2007a) and gave rise to the breakdown of the Gypsy and Traveller coalition in the early 2000s. Where often Gypsies and Travellers are bound together by their common experience of social exclusion, their cultural differences and competition over safe places to stop and stay on can lead to conflict. One GTAA stated, '*There seem to be strong negative perceptions of some of the other Travelling groups and a reluctance to be associated with* them' (NA2). Given the small number of legal stopping places for Gypsies and Travellers to live on, poor access to services, experiences of discrimination and the diminishing amount of seasonal work available to them (and competition for that work from Eastern European migrants) (Cemlyn *et al.* 2009), it is unsurprising that tensions should arise between the Gypsy and Traveller communities.

The GTAAs show that conflict between Gypsies and Travellers and the settled community is variable, with most conflict focusing around the development of new sites, unauthorised encampments or illegal developments. In the countryside, the conflict over such sites was found to be amplified by the rural nature of the environment. Despite the fact that Gypsies and Travellers are commonly placed within the context of countryside environments in the public and Gypsy and

Traveller imagination, the setting up of Gypsy and Traveller sites on green belt land has led to numerous conflicts, as has also been noted extensively in recent press surrounding the eviction of the Dale Farm site in Essex (Brown 2009). Notably, analysis of the GTAAs also found examples of peaceful co-existence of Gypsies and Travellers with settled communities where sites were legal or tolerated and long-standing.

Gypsies and Travellers rely heavily on unauthorised encampments, both roadside and tolerated, are commonly under eviction notice or fear of eviction, experience extensive racism and harassment, have poor relations with the settled community, lack access to and distrust services generally and feel discriminated against by them, and have an ambivalent relationship with the police. This set of circumstances that might lead us to describe Gypsies and Travellers as 'communities of shared risk' (Garland and Chakraborti 2004), result in a lack of trust and confidence between Gypsies and Travellers and the authorities responsible for managing and supporting them. The GTAAs outline the needs of Gypsies and Travellers in relation to accommodation and service provision to resolve a number of these poor circumstances. However, the type of site lived on makes the provision of services complex and difficult to deliver. For example, a family living on a settled, legal site may need access to GP services and education, whereas those living on the roadside need those services, but primarily seek clean water and basic sanitation.

Discussion

The research findings outlined above identify the problems and issues faced by Gypsy and Traveller communities. I have already considered in brief the framing of Gypsies and Travellers by control agencies as a problem requiring effective enforcement practice and it has been noted that such enforcement is more recently coupled with a supportive aim to provide services to Gypsies and Travellers who present welfare concerns. This placement of Gypsies and Travellers within a communitarian policy remit requires them to engage with agencies, as well as requiring agencies to engage with them beyond enforcement. As the previous chapter noted, Gypsies and Travellers are recognised as part of the wider, diverse community and particularly as part of a countryside idyll. However, their arrival in the countryside without formal spaces for them to live on poses problems for the authorities who perceive the Gypsies and Travellers as risky.

The authorities perceive Gypsies and Travellers as risky on three levels. Firstly, they pose a risk of disruption to the local community. As identified above, Gypsies and Travellers have poor relations with settled communities who are particularly concerned when unauthorised encampments arrive in their area. Community tensions cause social conflict that results in Gypsies and Travellers being subject to harassment and the authorities moving them on as rapidly as possible, rather than engaging with their support needs or considering where they may go. The

GTAA analysis shows that Gypsies and Travellers tend to stop and stay within a particular area and so multiple evictions act simply as a merry-go-round (Morris and Clements 2002). The failure of authorities to effectively engage with Gypsies and Travellers as part a diverse public is likely to increase the divide between them and the settled community and hence augment the social conflict. When such conflict occurs, the police utilise further enforcement measures and engage with a range of public order policing tactics to manage Gypsies and Travellers (James 2005) that further reduces their confidence in the police and authorities generally. Gypsies and Travellers respond to this by retreating into more hidden spaces (James 2007a) and thus do not embrace their responsibilities as required within the communitarian policy environment.

The ACPO guidance on managing unauthorised encampments, while recommending rapid eviction in some circumstances, does recognise the need to allow some unauthorised encampments more time and engagement from authorities. The guidance is written within the language of human rights discourse and the communitarian ideal, seeing responsibilities lying at the feet of Gypsies and Travellers, as well as the agencies that are required to engage with them. It is a typically practical police response to the issues raised by unauthorised encampments. As its architect, Bill Holland states, 'The guidance doesn't pretend to be this long-term answer. But I hope it will be seen as a pragmatic approach to the situation as it stands today' (Police Professional 2009: 3). The risk posed by Gypsies and Travellers as a cause of social conflict is potentially costly to authorities on a number of levels. The cost of policing public order problems is high, but additionally, the social cost of conflict is potentially higher as confidence of the settled population in the authorities is likely to be reduced if they do not remove Gypsies and Travellers from their area. As Morris and Clements (2002) have noted, the costs of managing Gypsies and Travellers are complex and multiple.

The second risk posed by Gypsies and Travellers to the authorities is the risk of high fiscal costs. Gypsies and Travellers, as has already been established, have very poor outcomes in relation to health and welfare. For the authorities to provide comprehensively to Gypsy and Traveller communities it would require the operation of numerous resources from multiple agencies in policing, local authorities, health, education, social services and employment services. Also, the longer that Gypsies and Travellers remain, the higher these costs are likely to be as needs are indentified and conflict with the settled community results in further service delivery requirements from multiple agencies as outlined above. Linking closely to this fiscal risk posed by Gypsies and Travellers is the third risk they pose to the authorities that is the perceived risk of crime associated with Gypsies and Travellers, beyond illegal encampment.

There is a perception within the settled community that Gypsies and Travellers, particularly those on the move, are involved in criminal activity (Clark and Greenfields 2006; James 2005; Power 2004). Evidence from research suggests that Gypsies and Travellers are actually no more likely to be criminal than the settled population (Dawson 2000), though it is extremely hard to make clear comparisons, given that ethnic monitoring within the criminal justice system tends not to include Gypsies and Travellers. There is more evidence that Gypsies and Travellers are victims of crime than perpetrators of crime. It is at this point that the perceptions of risk of Gypsies and Travellers and the authorities converge, as Gypsies and Travellers see their lives as risky on three levels, the first of which being the risk of victimisation they experience from the settled community, and, to some extent, from other Gypsy or Traveller communities. The failure of the authorities to address their needs as victims or their perception of the authorities as unwilling to help, and hence the low level of reporting of crime by Gypsy and Traveller communities, means that Gypsies and Travellers recognise their lives as risky enterprises, unsupported by the wider community and its agencies.

The failure of authorities to recognise Gypsies and Travellers' experiences of crime as victims exacerbates their lack of engagement with agencies and consequently their likelihood of living beyond the reach of provision. Gypsies and Travellers' lives are therefore risky in a second sense as they risk poor health, education and welfare by living nomadically and being unable to engage with the services that should be available to them. Finally, Gypsies and Travellers' lives are risky as they commonly live under threat of eviction from police or local authorities, who are keen to move them on in order to avoid the risks they pose to the authorities as described above. Thus, a vicious circle ensues, whereby the risks posed by Gypsies and Travellers to agencies largely determine the risky lifestyle they live that in turn augments authorities' perception of them as risky. As Gypsies and Travellers and the authorities perceive these risks differently, so the situation fails to be resolved, despite attempts at inclusion through the application of communitarian principles; Gypsies and Travellers are not provided with their rights and consequently are not prepared to accept their responsibilities.

Conclusion

The contested nature of risk results in a failure to provide for Gypsies and Travellers and their continued social and physical exclusion to the margins of society (James 2007a). This marginalisation is played out in countryside spaces, where Gypsies and Travellers are most likely to find locations to hide, but where their presence is also more easily noted by rural communities who question the presence of outsiders (Garland and Chakraborti 2004). The lack of provision for Gypsies and Travellers means that settled communities are unable to effectively engage with Gypsies and Travellers and so continue to fear them.

The relevance of the processes of inclusion/exclusion identified here are increasingly important internationally as Gypsies and Travellers are recognised as the largest ethnic group in the new Europe. Their experiences are reported in the international press and considered by organisations such as the Organisation for Security and Co-operation in Europe whose remit is to prevent, resolve and conflict, and the European Union Agency for Fundamental Rights which has identified policing Gypsies and Travellers as a problematic that should be addressed trans-nationally (James 2007b). In England, Government policy to assess the accommodation needs of Gypsies and Travellers via GTAAs has pushed forward the need to debate and discuss the inclusion of Gypsies and Travellers as part of the broad diversity of our communities. However, any resultant provision is slow in coming (Brown and Niner 2009) and being hard fought by settled communities (*The Herald* 2009). As a consequence tensions between Gypsies and Travellers and settled communities rise and the positive aims of inclusion are lost and assimilation or further exclusion expected.

Chapter 13

A Trip in the Country?
Policing Drug Use in Rural Settings

Adrian Barton, David Storey and Claire Palmer

Introduction

It is a feature of late modernity that sections of heavily urbanised societies such as Britain make much of the idyllic nature of 'the country' and the benefits of 'rural living'. Chief amongst those alleged benefits are the apparent problem-free nature of rural settings and the absence of many of the inherent social ills associated with city life. Although rural life is not without its problems, these are often different to those affecting cities (Department for Environment, Food and Rural Affairs 2000). Although we are all aware of the potential downsides of urban life – pollution, poverty, crime and anti-social behaviour – our determination to believe in the rural idyll means that we remain, largely, blissfully unaware of the existence of similar problems in rural locations. This is partly due to the tendency for rural problems to be under-explored by social scientists and under-reported by the media. It also stems from a misguided belief that 'rurality' is a holistic geography, when, as other chapters in this book indicate, this is clearly not the case.

There are a number of 'ruralities' and they suffer from many of the problems associated with the very worst urban environments. The difference is that these problems often manifest in different ways and, accordingly, are dealt with in different fashions. Drawing on empirical data, this chapter seeks to address these misconceptions of the idealised, problem-free, holistic nature of the rural idyll by focusing on illicit drug use in two rural settings. The chapter reprises some of the literature surrounding rurality, young people, drug use and the context in which it is policed. It then provides some empirical detail on drug use in two predominantly rural counties, Herefordshire and Cornwall. This is followed by some observations on the manner in which drug-taking is policed within the two counties. The chapter concludes with a discussion regarding the use of, and responses to, illicit drug use in rural settings.

Multiple Ruralities

For the purposes of this chapter, rural is taken to refer to areas with dispersed populations and characterised by villages and small towns. However, we are very aware that it is now generally accepted that there is no one single version of the 'rural'. Economic and social heterogeneity have created a multiplicity of overlapping social spheres within rural space (Mormont 1990). While idyllic constructions of the rural hold a very definite resonance for many, the reality of life for some sections of the rural populace is rather different as evidenced by rural poverty, deprivation, homelessness and other forms of marginalisation (Cloke *et al.* 1995; Cloke *et al.* 2001; Cloke and Little 1997; Philo 1992). Yet, in line with broader constructions of the rural, there has been a tendency to portray small towns, villages and the countryside as offering very pleasant and safe environments for children and young people, where they can enjoy happy and healthy lifestyles (Jones 1997).

The reality is of course much more complex but it is only quite recently that some researchers have focused attention on young people and their experiences of rural or small town life (see for example Flood-Page *et al.* 2000; Jentsch and Shucksmith 2004; *Journal of Rural Studies* 2002; McGrath 2001; Shucksmith 2004). Many young people in rural areas have to contend with low pay; poor working conditions; limited choice and availability of work; worries about future employment prospects; and highly constrained recreational opportunities (Matthews *et al.* 2000). Poor transport links and service provision may serve to compound these problems, contributing to heightened feelings of isolation and alienation. It is also argued that the lack of anonymity in rural areas has consequences for rural youth. Davidson *et al.* (1997) suggest that the close-knit character of rural communities means that young people have a heightened social visibility and perceive themselves to be almost constantly under observation (Matthews *et al.* 2000). The methodological issues associated with accessing directly information on young lifestyles further compounds our relative ignorance (Leyshon 2002). All of this points to a disjuncture between perception and reality, with popular visions of idyllic rural communities existing alongside processes of social exclusion.

Illicit Drug Use in Contemporary England and Wales

One of the most consistent sources on illicit drug use has been the British Crime Survey (BCS), now conducted on an annual basis with members of the public in England and Wales. As part of the survey, respondents are asked to state whether they have used illegal drugs in their lifetime and/or in the year prior to the survey. This section, which is self-administered and confidential, uses a number of 'check questions' designed to rule out exaggeration or misrepresentation.

The findings suggest that rates of illicit drug use in England and Wales appear high: over one third of the population (around 11 million people) aged 16–59 claim to have used an illicit substance in their lifetime (Chivite-Matthews *et al.* 2005; Hoare 2009). However, 'lifetime' is something of a misnomer as highest usage tends to occur amongst young adults, with rates tailing off in later life. Those in the 20–24 age range are twice as likely to use an illicit substance as the rest of the population with one quarter of 16–24 year olds claiming to have used cannabis in the year prior to the survey (Chivite-Matthews *et al.* 2005). Men were twice as likely as women to have used illicit drugs (Hoare 2009).

The most recent sweep of the BCS (Hoare 2009) revealed some surprising variations between drug use in the country and city (Table 13.1). Although (self) reported drug use is higher amongst adults in urban places, it appears that usage amongst young adults (16–24) is proportionally highest in rural places. Illegal drug use may not only be seen as part of many young people's lifestyles but, significantly, part of the lifestyles of young people living in rural areas.

Table 13.1   Urban and rural differences in declared illicit drug use

	Proportion of 16 to 59 year olds reporting use of drugs in the last year	**Proportion of 16 to 24 year olds reporting use of drugs in the last year**
Urban	10.7%	22.4%
Rural	7.3%	23.5%

Source: Hoare 2009

Despite this, knowledge of drug use by young people in the countryside is extremely limited, rendering it difficult to discuss it in any meaningful detail (Brown and Young 1995; Davidson *et al.* 1997). Drug-use is still considered by many academics and policy makers to be an issue confined largely to urban places (Henderson 1998 but see Cocklin *et al.* 1999). Consequently, little is known about drug-use in rural localities and how it is policed. Arguably, more locality-based studies are needed to examine the complex interactions that occur between users, dealers, the police and other agencies in order to determine how, how often and where drugs are used in the countryside.

The Drugs Situation in Herefordshire and Cornwall

To begin filling this gap, this chapter goes on to describe two studies of drug-use undertaken in Cornwall and Herefordshire by the authors. These were conducted independently (Barton 2005; Storey and Palmer 2002) and, while not intending to be directly comparable, provide different, but complementary, perspectives on rural drugs. Storey and Palmer's work examined drug-taking from the perspective of users and a drugs advisory service; Barton's examined police efforts to identify and disrupt the supply of goods. Both pieces of work made reference to the incidence of drug-use in their respective areas, providing some much needed local detail on the BCS's national picture. The following sections introduce the rural characteristics of these places and go on to describe drug-use within them.

Consideration is then given to the policing of drug use by various agencies in Cornwall and Herefordshire.

Leominster, Herefordshire

In many ways Herefordshire encapsulates many peoples' notions of the rural, problem free idyll: 'most people, when they think of Herefordshire, imagine a beautiful and largely rural county, nestling between the Welsh and Malvern Hills, far from the bustle of big city life – a rural idyll' (Hereford Conservative Party election candidate, Virginia Taylor, *Hereford Times*, 3 April 2003). However, the same speaker goes on to say: 'in many respects of course it is. But scratch the surface and you will find a cancerous growth: drugs – and particularly hard drugs'.

The problematic nature of illicit drug use in Herefordshire was recognised over 20 years ago through the creation of a drug advisory service (known as DASH, funded and managed through the NHS Herefordshire Primary Care Trust). DASH coordinates treatment for drug dependant users and its work includes prescribing, key working, needle exchange, GP liaison, counselling, relapse preventions, acupuncture and outreach. It offers a free and confidential service to drug users, families and friends and it liaises closely with other related services and agencies.

As Herefordshire is a rural county with a widely dispersed population it proved difficult for those living outside Hereford to access drug services. For this reason, the idea of outreach provision was considered a viable option to access drug users in the smaller market towns, and offer them a 'bridging' service, as well as support in their own communities. One of these initiatives was located in Leominster, a market town of some 11,000 people that serves a large rural hinterland. It was estimated that, between 2000 and 2002, between 15 and 30 people were utilising their services at any one time, the majority of them young adult males.

As part of an evaluation, interviews were conducted with drug-users that were clients of the project (Storey and Palmer 2002). Although these were perhaps only a fraction of the area's drug-users, the interviews offered opportunities of gaining insights into drug-taking in small towns and rural areas from the perspective of users themselves. These interviews suggested that illicit drugs in Leominster had become more accessible to younger people. The survey revealed that people are taking drugs at a relatively early age and many graduated to heroin in their mid-teen years. Nineteen drug users had tried both types of cocaine but the most regularly used drugs were cannabinoids (in line with national figures), with 10 of those interviewed using them daily. Twenty-two of the drug users interviewed used heroin, ecstasy and alcohol (Figure 13.1) though only four used heroin daily and five of those who used it had not taken it in the past three months.

BCS data suggest that around 8% of those aged 16–24 use Class A drugs, a figure which has remained relatively stable for some time. In a group of young people interviewed in Leominster, one third claimed to have taken illicit drugs and

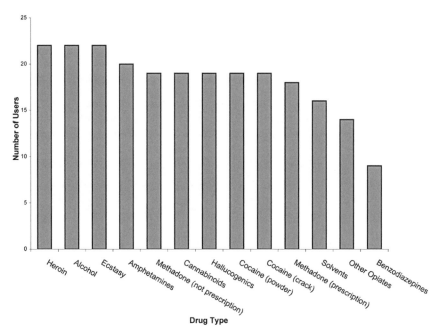

Figure 13.1   Drugs used in Leominster by outreach clients

the vast majority claimed to know at least one person who took drugs. While Class A drugs (ecstasy, cocaine, crack, heroin, LSD, magic mushrooms, methadone) were not being abused in large numbers, over a quarter of those interviewed had smoked cannabis and six had used solvents (Figure 13.2). The popularity of cannabis is hardly surprising given its relative ease of accessibility (Barnard and Forsyth 1998). Leominster usage rates for all other drugs are substantially lower, once again reflecting national trends (Condon and Smith 2003).

Survey results estimated that there were approximately 53 heroin users in the town and its surrounds consuming drugs within their own 'using' circles. While users within the circle knew of each other's identity due to using the same dealers and spaces, their taking and dealing activities remained relatively isolated from other users in other 'circles' and from the town's population at large, thereby contributing to the hidden nature of the phenomenon and reflecting a particular geography of use.

A number of drug users suggested that the availability of drugs is 'like a tap, it can either flow, drip or totally dry up'. Almost all of the users interviewed indicated that heroin was very easy to obtain clearly reflecting the network of associates which drug users have. The majority of the interviewees also suggested it was quite easy to get methadone illegally. Almost half of the participants in the young person's survey claimed to have been offered drugs at some point in Leominster (although only 10 admitted to having actually purchased drugs in the town)

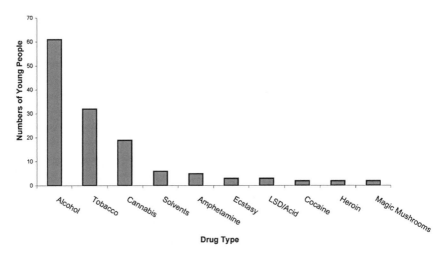

Figure 13.2   Drugs used by young people in Leominster

reflecting the levels reported by Barnard and Forsyth (1998) in rural Scotland. The majority of young people felt that alcohol, tobacco, solvents and cannabis were very easily available. While harder drugs were seen as more difficult to obtain, there was still a number of young people who felt that these drugs were relatively easy to acquire. There is an obvious distinction between those who are likely to have regular contact with dealers and other drug users and those who do not, but there is, nevertheless, an awareness of drug availability on the part of some young people. User circles clearly exist (Cragg Ross Dawson 2003) and those operating in them know where to obtain their supplies in contrast to what can be seen as recreational users for whom supply may be more difficult (Parker *et al.* 2002).

Cornwall

Cornwall is a county that also conjures up ideas of a rural retreat. With a population of around 500,000, it is internationally known for its rugged coastline and beautiful beaches. However, it too has a number of problems associated with illicit drug use. A 2004 audit of drug services noted that there were around 2,500 problematic drug users in the county (around 5% of the population), of which about 55% were in treatment at any given time (Barton 2005). In addition, the same report provides evidence taken from police data, which implies a thriving recreational drug culture within the county. Most of the recreational drug use centred around the key tourist towns of Newquay, Truro, Penzance and St Ives, but overall drug use was spread evenly between the rural and urban areas of the county. It must also be noted that whilst cannabis is absent from the list of seizures, police data reports that cannabis use is widespread throughout all major towns.

Table 13.2   Drug seizures in Cornwall 2003–04

Type of drug and amount seized	County-wide
Heroin	788.24g
Cocaine	136.40g
Crack cocaine	29.16g
Ecstasy	12,713 tabs

Source: Barton 2005

In terms of drug dealing and supply networks, Cornwall has a varied and almost cosmopolitan situation. There are two factors attached to this. The first is that because of the isolated nature of most of its coastline, Cornwall is often a landing point for drugs being smuggled into the UK. Once in the county, these large shipments are then transported to major cities such as Bristol, Liverpool and Manchester for national distribution. Ironically, they are often transported back to the county to be sold in the major towns. It is also the case that dealing networks are often spread between adjacent towns, as detailed in Figure 13.3 (derived from police intelligence). Clearly, drugs are being imported into the county for distribution but drugs are also being re-sent to dealers outside of Cornwall.

As a result of what is clearly an entrenched and thriving drug culture, and mirroring developments in Herefordshire, there are a number of drug-related services in Cornwall. These are a mix of third sector voluntary groups that tend

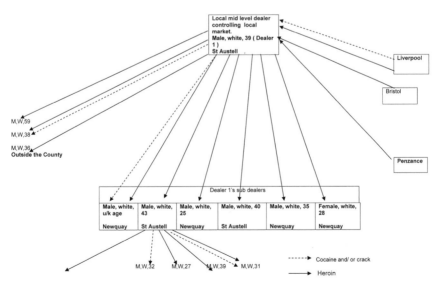

Figure 13.3   The geographies of drug dealing

to be either town- or district-based. It is these groups that provide much in the way of outreach and drop-in services. These third sector groups have varying levels of funding and operate according to a number of ideologies. There are also statutory services such as the DAAT supplying countywide coverage and, in the main, providing structured treatments. Clearly then, both counties have what would appear to be significant problems with illicit drugs with a range of agencies involved in responding to these. The following section examines efforts to police these by considering, firstly, the nature of policing at the national level and, secondly, the impact of local policing projects in Herefordshire and Cornwall.

Policing Drug Use

National Perspectives

For many people policing revolves around the images provided on an almost daily basis by television programmes like *The Bill*. However, Reiner (2000: 1) argues that:

> It is important to distinguish between the ideas of 'police' and 'policing'. 'Police' refers to a particular form of social institution, while 'policing' implies a set of processes with specific social functions.

He continues, noting that 'policing' is, in fact, a set of activities that attempt to maintain a particular form of social order. That order may result from a society wide consensus, or it may be the result of latent conflict between differently placed social groups. Whatever the case, the function of 'policing' is to ensure that a particular form of social order is encouraged, maintained and upheld. For our purposes then, 'policing' can be seen as: '… systems of surveillance coupled with the threat of sanctions for discovered deviance' (Reiner 2000: 3). This definition can encompass the work of not only the police and the other law enforcement agencies such as Customs and Excise, but can also be seen to cover much of the work of what are defined as 'welfarist' agencies, including social workers, probation officers and the medical profession. Indeed, policing in some cases can be carried out by peer groups and family.

Using this definition allows us to conceive of 'policing' as being on a continuum at the 'hard-end' of which are the law and order agencies, employing legislation, often a blunt instrument, to achieve control and conformity. At the other 'soft-end' are the 'social' agencies, which often employ less overt, more subtle methods, but nevertheless use the twin tools of surveillance and threat of sanctions as a means of controlling populations, allowing these agencies to be characterised by some as 'iron fists in velvet gloves' (George and Wilding 1992). This is important as the reality of most drug users' lives is that their contact with 'the police' is sporadic and confrontational whereas their contact with socio-medical and welfare agencies may be more regular and based around support and help.

In the past the monitoring and policing of illicit drugs tended to be seen as the responsibility of the police with a primary emphasis on law enforcement. However, roles have changed somewhat in recent decades such that education about the dangers of drug (mis)use and efforts at minimising or reducing the harmful impacts on drug users, local communities and wider society are now seen to be integral components of a broader view of policing drugs. Intrinsic to this are such things as the referral of users to appropriate support services and facilities such as needle exchange and methadone programmes (Davidson *et al.* 1997). This broader approach stresses inter-agency collaboration aimed at prevention and protection, rather than simply the pursuit and prosecution of those involved.

The UK government's drugs strategy encompasses both efforts to reduce drug supply as well as the treatment of users and the education of the public, particularly younger people, about the problems of drug use. The creation of Drug Action Teams (now commonly subsumed within Drug and Alcohol Action Teams) reflects this shift. DAATs are partnerships aimed at coordinating the activities of relevant agencies at a local level and they work closely with (or have been merged into) Crime and Disorder Reduction Partnerships or Community Safety Partnerships. The DAATs are thus a key element in the implementation of the guiding principles of current government strategy – enforcement, treatment, protection, education, and social services (HM Government 2008). In general terms then the policing of drugs can be seen to retain the traditional vision of detection and interruption of supply combined with a robust strategy aimed at changing drug users behaviour via Drug Intervention Programmes and the use of Drug Treatment and Testing Orders which 'force' offenders to engage with treatment. Alongside this there is the utilisation of multi-agency partnership (in tandem with health, probation and other services) aimed at somewhat broader ends associated with protection and prevention.

The utilisation of an inter-agency approach emphasising partnership, locality and community can also be seen to reflect broader policy trends in relation to the governance of both rural and urban areas (Storey 2010). The precise nature, composition and performance of governance are influenced by local power relationships in different places (Woods and Goodwin 2003). Thus, a *geographical* understanding of governance is crucial. Goodwin (1998) maintains that governmental structures become 'sites of regulation' that contribute to the uneven development and influence of specific modes of regulation. It is therefore important to examine the interpretation and delivery of policy by particular agencies and how these actions impact spatially on specific modes of regulation (Cloke and Goodwin 1992; Goodwin and Painter 1996). As Woods and Goodwin (2006) point out, the ways that these agencies combine can vary considerably over space, reflecting local variations in the way that power is negotiated and exercised. Returning to Cornwall and Herefordshire, the next section starts to examine how multi-agency policing, in its broadest sense, plays out in particular localities.

Policing Responses in Herefordshire and Cornwall

It is clear that in Herefordshire and Cornwall there is a 'drug problem' and, by extension, this needs to be 'policed'. Equally 'policing' is not simply about the actions of the police but, instead, it takes place on a continuum and is conducted by both formal and informal agents so that much 'policing' is done by socio-medical and welfare-based agencies. Indeed both sets of data revealed similar issues.

Drug users in both areas were more concerned with the actions of welfare agencies than those of the police. Drug users interviewed had little to say about the 'policing' of their drug use by the police. Clearly, there is an awareness that the police are there and are able to disrupt supply, making drugs more difficult to obtain at certain times. The police also have the ability to arrest and detain drug users for both the primary crimes of possession and supply and the secondary crimes of stealing in order to fund drug purchases. Even then, the intimate local knowledge and tight-knit nature of the drug using populations in rural areas seemingly negates the need to buy drugs from 'strangers', making arrest for possession and supply a 'low risk'. Beyond these points the police seem to be peripheral to drug users' lives.

Equally, looking at rural drug use from a police perspective, it is clear that the police have considerable intelligence on the dealer and user networks operating within the counties to the extent that, arguably, they are in the position to make considerable in-roads into disrupting the supply. Periodic drug raids in Leominster and elsewhere, and actions such as the searching of vehicles suspected of transporting drugs in Gloucestershire and Herefordshire are prime examples of this type of policing (*Hereford Times*, 7 December 2007). The extent and apparent depth of this intelligence could be attributable to the high levels of surveillance and local knowledge afforded by rural living as noted earlier. Nevertheless, it could be suggested that the police are seemingly content to monitor and only partially disrupt supply. This may be because the illegal drug market mirrors the other, licit, markets in rural locations: that is, small- to medium-sized businesses that serve the needs of the local population and knowledgeable of local culture and mores. To put it another way, intensive interference in the drug market by the police would open the door for outside organisations who would be more remote, unknown, and potentially disrupting and dangerous. Maintaining the status quo may be the favoured approach as it enables them to retain detailed local knowledge about who is using and selling and thus retain an element of control and stability which is often lacking in urban drug markets (Caulkins and Reuter 1998).

Drug users in Herefordshire and Cornwall had a lot to say about the socio-medical and welfare agencies and it is here, we would argue, that the bulk of 'policing' rural drug use occurs and as a result, it is at the socio-medical and welfare aspect of drug use where the discrepancy and difference occurs. Just to re-cap, above we defined 'policing' as 'systems of surveillance coupled with the threat of sanctions for discovered deviance' (Reiner 2000: 3) and drug agencies have elements of this definition of policing embedded within their work. In Herefordshire the county

DAT evolved into a Community Safety and Drugs partnership and was re-branded as Safer Herefordshire in 2009. One aim of its crime, disorder and drug reduction strategy is to reduce the harm caused by illicit drug use. Overall, socio-medical and welfare based agencies (such as DASH) are far more likely to see recreational drug users than the police and even the most chaotic problem drug user will be in contact with their key drug worker more than any police officer.

Work conducted in Cornwall (Barton 2005) provides an interesting look at some of the problems faced by agencies and their clients in policing drug use in a rural setting. These include problems in access and provision of services across the county due to poor public transport networks. General Practitioners were perceived to be generally ill-informed about addictions and the provision of relevant services and were seen as reluctant to provide or refer for treatment. This presented the service user with difficulties in getting a diagnosis or finding out where to go for help. There was general agreement that, given the dispersed nature of the users and the difficulties in accessing these services, a 'one stop shop' for initial contact, would be very helpful. This would need widespread publicity and interagency cooperation, but it would reduce confusion and delay while providing equality for referral and access to services. It was felt that there was a lack of resources and facilities for young people and more attention should be focussed on education and harm reduction. Inter-agency cooperation was seen as another key issue, with rivalry between the agencies fuelled by funding problems causing lack of feedback on referrals, lack of flexibility, choice, coordination, continuity, fragmentation and communication. Where supported housing schemes are available the feedback was extremely positive, but it is clear that these vital services are rationed and scarce, and more would be appreciated. It seems, for some drug users, that the transition from detox to independence is too abrupt and there is little help available at times of difficulty and stress. For those coming out of prison suitable affordable housing seems impossible to attain with some feeling stranded in hostel accommodation following discharge from detox/prison/rehab. There are many problems for the homeless in accessing help and treatment as they often do not have a contact address for appointments. Another area of concern was the lack of detox beds for both alcohol and drugs in Cornwall, causing difficulties in accessing them, delays and waiting lists.

More specifically rural problems revolve around transport and the difficulties infrequent and sporadic services create in terms of access and availability, relatively low numbers of users leading to limited provision. For example, whilst there is clearly a need for out of hours services and drop-in centres there is little point in providing these if there is no corresponding bus service which allows users in outlying areas access. As a result, service provision tends to be patchy and only available at very restricted times. Equally, relatively low numbers means that funding is low and competition between (and in some cases) across agencies is fierce leading to a reluctance to share.

Evidence suggests high levels of satisfaction with welfare-based agencies, but this is tempered with a realisation that, arguably as a direct result of the rural nature of the settings, services are limited and constrained. For example, the Leominster outreach project has managed to reach a sizeable proportion of the town's drug users and has clearly been a success in terms of supporting users and offering counselling/referral services as well as practical advice, but it is stretched in terms of time, resources and personnel. This means that the serious drug issues faced by some young people in Leominster are almost certainly going unnoticed. It is also worrying that less than half of the young people who responded to the questionnaires had actually heard of the outreach project. For some it was seen simply as somewhere that dealt with 'druggies' and people experiencing heroin addiction. In much the same way, drug users in Cornwall were supportive of the services they used but were acutely aware of the limitations. Many of the perceived gaps in service provision were seen as a direct result of the rural nature of the county: lack of transport; difficulty in accessing services out of hours; poor information and poor understanding of drug users' needs; and lack of affordable housing.

Conclusions

It is clear that illicit drug use occurs in rural areas and creates a range of problems for users, wider communities and for those charged with curbing, controlling and policing the phenomenon. Any vision of 'policing' drugs in rural settings that only concentrates on the work of the police is, at best, partial. Our work indicates that much of the 'policing' work is carried out by socio-medical and welfare agencies with some degree of success. We would also argue that policing rural drug use carries with it a unique set of opportunities and constraints resulting from aspects of rurality alluded to earlier. In terms of opportunities the comparative lack of mobility, limited opportunity for anonymity and high visibility that living in close-knit relatively small communities affords agencies and individuals allows high levels of detailed knowledge to be gained which can allow police to monitor and in some instances 'control' local drug markets bringing a degree of stability to what, in urban settings, are often violent and chaotic arenas. Equally the socio-medical and welfare agencies have the same degree of knowledge and can tailor responses to fit changing local conditions. Conversely, high visibility, limited access to public transport, cheap housing and specialist facilities, and the apparent lack of highly coordinated service provision mean that rural drug users, especially those who are problematic or in need of specialist services, often find that the range and depth of services available to them are below that of their urban counterparts. It is ironic that for drug user and non-drug user alike being part of rural life means learning to live with a reduced level of service provision.

Chapter 14

'It's Not All Heartbeat You Know': Policing Domestic Violence in Rural Areas

Greta Squire and Aisha Gill

Introduction

Three million women across the UK experience rape, domestic violence, forced marriage, stalking, sexual exploitation and trafficking, female genital mutilation or crimes in the name of 'honour' each year (EVAW 2008). Individual women and girls count the cost in terms of cuts and bruises, broken bones, miscarriages, sexually transmitted diseases, long-term mental health problems, substance abuse and social exclusion, and even, in extreme cases, death. The cost to society to respond to these crimes is over £40 billion a year in England and Wales alone. Gender-based violence is one of the most serious inequalities facing women and girls in the UK (EVAW 2008). All of these crimes are justified or excused by myths and stereotypes; all are supported by the dynamics of power and control; all suffer from high levels of under-reporting and extremely low conviction rates. There is also a widespread tendency towards repeat victimisation, and there are long-term social, psychological and economic consequences for victims. Above all, the state has historically failed to tackle the crime of gender-based violence (EVAW 2008).

One form of gender-based domestic violence is proportionately nearly as common in rural areas as it is in urban or inner-city settings (Walby and Allen 2004; Women's Institute 2008). Research suggests that although incidence of this offence is statistically higher in inner-city and urban areas, one in four victims of domestic violence lives in a rural location (Walby and Allen 2004). Perhaps inevitably, the higher rate of incidence of domestic violence in urban areas justifies the proliferation of services and resources in these communities, but this should not be an excuse for ignoring rural victims or failing to provide adequate services for them. Rural domestic violence, like rural racism, is decidedly an 'invisible' problem (Chakraborti and Garland 2003).

McCarry and Williamson (2009) contend that there is a tendency to overlook the plight of rural victims and to overstate the similarities between the problems of rural women and those of their urban counterparts, leading to serious misunderstandings, policy neglect and injustice. Rural domestic violence should be treated as seriously as urban domestic violence yet, at the same time, it should be treated as a unique phenomenon. Rural communities are manifestly different from urban communities and so rural domestic violence against women has its own parameters and features.

This chapter will begin by defining what is meant by 'rurality' and will then turn to examine the particular difficulties faced by women experiencing domestic violence within rural areas. Following this discussion, the chapter will examine the police response to domestic violence, and in particular the services female victims living in rural areas receive. The chapter concludes with a discussion of some recent exploratory findings, and presents tentative recommendations about the route policy needs to take if domestic violence in rural areas is ever to be genuinely addressed.

Defining Rurality and Rural Communities

The term 'rurality', like many key concepts within the social sciences, remains somewhat loosely defined. This discussion draws heavily on postmodern interpretations of power and class. Specifically, recent commentators have attempted to deconstruct existing meanings of rural areas that over-privilege the landowning gentry. Consequently, recent attempts to define or describe rurality rely less on differentiating the rural from the urban (based on quantifiable traits such as population figures or geographical proximity) (Hoggart 1990) and more on defining the common characteristics found in communities described as 'rural' (Mormont 1990). While smallness and isolation are traits essential to the character of the rural community, Mormont maintains that there is no quantifiable point at which a community becomes too large or not isolated enough to qualify as rural. In other words, it is the *internal* features of a community that define it as rural, rather than its size or proximity to urban areas. Rural communities have cultural characteristics that are based in tradition or history, not simply geo-spatial circumstance, that give rural communities a sense of longstanding continuity. These are reinforced by a unifying structure, either institutional or economic in nature, that continues to tie the community together (Mormont 1990). These defining features of rural communities distinguish them from urban societies and demand that they be treated distinctly.

It is in this context that some feminists have suggested that urban-based feminism does not address the realities of women's lives in rural communities (Alsop *et al.* 2002; Little and Panelli 2003; Panelli *et al.* 2004). We would go one step further and argue that a wider problem within feminist thinking is that disparate and sometimes conflicting voices within feminism (Tong 1996) can lead to incoherent policy recommendations. This is reflected, for instance, in the criticism of mainstream white feminists by black feminists in the late 1970s and early 1980s (Williams 1989). The problem is further compounded because feminist social theories of rurality fail to address the significance of gender within rural space adequately. This leaves rural women with no unified theory that accounts for their roles as women and as members of a rural community (Little and Panelli

2003). This chapter is, then, a call to arms, a demand for a unified conception of rural communities that can promote the cause of female victims of domestic violence.

Domestic Violence and Rural Women

All studies agree that domestic violence remains indisputably significant in the UK. While the broad scope of the problem is well documented, there is very little research which captures the experiences of violence and abuse experienced by women in rural areas (Bradshaw *et al.* 2006; McCarry and Williamson 2009; Websdale 1995a, 1995b). The difficulty is that domestic violence is chronically underreported, particularly in rural communities where there is less privacy for domestic violence survivors.

Despite the lack of quantifiable data, there is a general consensus that the challenges facing survivors of domestic violence are in many ways made worse if they reside in a rural community. To take one example, escape or flight from abuse is much harder due to the physical and social isolation and the distance to or even unavailability of services (Coy *et al.* 2009; McCarry and Williamson 2009). The physical isolation that comes from living in rural communities increases the instances of violence against women, simultaneously limiting a victim's access to domestic violence services and prolonging their suffering (Websdale 1998).

Economic powerlessness and cultural mores make it difficult for victims to approach neighbours for help so that rural victims may remain in abusive relationships. It is clear that there are more barriers to reporting, seeking care for and prosecuting cases of domestic violence in rural communities than in urban areas, pointing to the possibility that rural forms of violence against women may be more difficult to combat. Websdale (1998) also found that rural communities are typically conservative and so resist acknowledging and dealing with social problems such as domestic violence. So-called 'close-knit' communities may even close ranks *against* victims, principally because people prefer victim-blaming to public scandal that brings shame on the entire community (Mackay 2000).

Very often the perpetrator of domestic abuse is a prominent or respected member of the community even though domestic violence is not specific to any particular socio-economic group (Walby and Allen 2004) and can happen across the classes. In rural areas there is a greater chance that a perpetrator will be a known figure in the community, and this undeniably affects the victim's decision whether or not to come forward. Above all, it is important to recognise that the degree of isolation experienced by women is directly linked to the degree of community awareness of the problem and the support offered to those who are abused, and that this in turn influences how long women remain in violent and abusive relationships. And it is clear that in these crucial areas – awareness and support – those rural communities fail abused women, and especially those from black and minority ethnic (BME) backgrounds.

Isolation is an even greater problem for BME women, as their ethnicity is commonly a factor in preventing them from accessing statutory services (Rai and Thiara 1997; see also Barclay *et al.*, this volume). And all female victims of abuse are at further risk because of the lack of information about domestic violence within their communities, and because of the absence of meaningful dialogue about effective intervention strategies (Gill and Rehman 2004). The absence of secure social networks in rural communities, where women are often severed from their extended families, only worsens this situation. This is especially true for (im)migrant women experiencing domestic violence, who have limited access to family support. Crucially, women from BME backgrounds who lack social networks are typically completely dependent on their husbands for their social, financial and emotional needs, but their husbands often give them inaccurate information concerning their legal and immigration status. Moreover, such women are unable to seek advice either from friends or professionals about how best to deal with their abusive situation (Gill and Rehman 2004).

Unfortunately, much of this must remain conjectural, because empirical research on violence against women in rural areas is insufficient, especially in the UK. There is clearly a pressing need for further work to be done in this area.

Policing Domestic Violence in Rural Areas

When abuse finally comes to light, how do the police respond? The reality is that this is not a matter of straightforward, old-fashioned policing. Intervention to stop and prevent domestic violence commonly involves professionals working in a number of different statutory and voluntary agencies. But the police occupy a unique place in society: they are effectively ordinary citizens who are granted extraordinary powers (Bowling and Phillips 2002; Newburn 2003; Reiner 2002). Their ability to change people's lives instantly means that they deserve particular scrutiny in cases of domestic violence. High-profile domestic violence incidents, like those of Julia Pemberton and, more recently, Banaz Mahmod and Sabina Akhtar, call into question how the police should act in domestic violence cases (Gill 2009). Policing domestic violence can be particularly difficult because the perpetrator is typically 'a loved one' and, for this reason, many officers have typically not wanted to get involved (Edwards 1989).

In the 1980s, the police were accused of not taking incidents seriously, particularly as domestic violence cases have been treated as civil rather than criminal matters, and of using inadequate recording practices that obscured the true scale of domestic violence. During the early 1990s, the Home Office worked alongside the police service to introduce measures to respond to these criticisms. Home Office Circular 60/1990 emphasised:

- the need for policy documents and clear strategies for dealing with domestic violence incidents;

- an interventionist approach based on the presumption of arrest when an offence has been committed;
- a recording process for 'domestic' incidents which reflects procedures for other violent crimes; and
- the establishment of dedicated units or specialist officers to deal with domestic violence incidents (Morley and Mullender 1994).

The attitude of not treating domestic violence as a serious crime has been tacitly rejected (Edwards 1989) and the police have therefore policies aimed at changing the handling of domestic violence crimes. Spurred on by public concern and the work of female advocacy groups, the police have responded positively: pro-arrest policies have increased the volume and altered the type of cases that come to the courts in relation to domestic violence. And yet, although the police are central to the handling of this offence, gaps remain in their ability to protect properly victims, particularly in rural areas.

Researchers and policy-makers have welcomed these developments, but nevertheless may have overlooked the impact that old, long-standing attitudes may continue to have. They too readily assume that police officers have embraced these more recent interpretations of domestic violence and that consequently they are entirely receptive to new legislation. Although most forces have introduced policies to deal with these needs, the translation of policy into practice had been less successful (Grace 1995). More recently, the need for a more effective and integrated response, especially in rural areas, has recently been stressed by the End Violence Against Women Coalition and other campaign groups (EVAW 2008).

Yet little is known about how the police response to violence against women in rural areas, and in particular their attitudes to this form of abuse. Although the actions of police officers in domestic violence cases has been observed *in general* (e.g. Hanmer and Griffiths 2000), little work has been done on how officers respond to domestic violence in *relation to their location*. This means that the impact that rurality has on police response has gone largely unexplored. The question then remains: how does a victim's locale affect the kind of police support she will receive? To begin answering this, we will now turn to field observations of the policing of domestic violence made by the first author in rural Oxfordshire, UK in order to provide a snapshot of policy dilemmas and shortfalls in the policing of domestic abuse in rural areas.

Policing of a Rural Area – Evidence from an Exploratory Study

It is important to state that the evidence provided here emerged from an exploratory case study. As Easterbrook *et al.* point out, 'exploratory case studies are useful when it is not yet clear which phenomena is clear and which is not' (Easterbrook *et al.* 2005). In this research the 'case' being examined is the work of the Thames Valley police force based in Oxfordshire, a county in the South East of England with the main population based in the city of Oxford.

For this chapter, the investigation relied upon documentary analysis of secondary literature and informal interviews with police officers, including representatives from the Oxford City Domestic Violence Unit, the Public Protection Officer responsible for Partnership Planning of Violent Crime in the county, and the county-wide Domestic Violence Co-ordinator for Oxfordshire. These interviews were semi-structured and lasted between one and three hours.

The following sections assess how the police within the Thames Valley Force are organised to respond to domestic violence; what services are available to women and children within Oxfordshire trying to escape from violence; the role of partnerships and how they function within a rural setting; and asking what – if any – problems arise from geographical constraints to partnership working?

Police Procedure within the County

The police within Oxfordshire respond to domestic violence through their specialist Domestic Violence Unit, based within Oxford. The attending officer typically responds to any incident of domestic violence, completing a risk assessment of the victim during a visit to the location of the abuse. This information then leads to a decision about what level of intervention is required. The risk-assessment tool used within the Thames Valley area is based upon the SPECCS (Separation, Pregnancy, Escalation, Cultural Issues, Stalking and Sexual Assault) evaluation (see Richards 2003), an important method used by the police to calculate risk and need. This approach is not dependent upon where the victim lives, but is completed wherever the incident is said to have taken place. Risk assessment is considered the central instrument for establishing the needs of domestic violence victims, and as such is viewed as a vital tool for the police to address this problem (Gamble 2005). This is particularly relevant within rural contexts, as access to services dealing with domestic violence is, as discussed above, more limited in rural areas.

The data gathered from this assessment then forms the basis for the involvement of the specialist Domestic Violence Unit (DVU). During the interviews it emerged that an officer from the DVU will only visit the victim if they are regarded as medium or high risk, as defined through the SPECCS criteria. From this information a package of services is developed, which often incorporates a multi-agency approach to dealing with the needs of the victim. Indeed, the two officers interviewed for this paper considered that their role was about the management of risk within a multi-agency context (Crawford *et al.* 2005), rather than something that was the sole domain of the police. They saw themselves as a 'first point of contact', and following the risk assessments other agencies would typically be brought in to supply relevant service provision, dependent upon need.

In Oxfordshire, alongside the DVU there is the Independent Domestic Violence Advocates (IDVA); there is one IDVA in the north of the county and one within the city, but they do not have dedicated IDVAs in either the South or West. The Domestic Violence Reduction Strategy for 2007–10 discusses the possibility that the IDVA programme might be extended to incorporate the rural areas, but there are no concrete details or funding plans for this, and no date has been suggested.

Service Provision within the County

As discussed earlier, in order to provide services to deal with the complex needs of victims of domestic violence (Crawford *et al.* 2005) there is a need for 'pluralisation', a multi-agency approach that includes, but is not limited to, the police, and involves a variety of agencies which are responsible for maintaining law, order and public safety (Yarwood 2005). With regard to Oxfordshire, this approach is exemplified by the work of the crime safety officers that respond to issues such as the replacement of doors broken during an assault. Like many advertised services, it is clear that crime safety officers are more accessible in the city than in rural districts. The Domestic Violence Unit Police Officer for Oxfordshire put this down to:

> A more energetic crime safety officer within the city compared with the rural districts, who was happier going cap in hand to chase money. (Domestic Violence Unit Officer)

Similarly, the funding of the Crime and Disorder Reduction Partnerships (CDRPs), which support such schemes, is also weighted toward the city, with rural areas even having to forgo such schemes entirely. Indeed, all of the interviewees indicated that access to services is more difficult in rural areas. One such service is the Specialist Domestic Violence Court; as the recent Home Office report states: 'the Court systems provide a specialised way of dealing with domestic violence cases in a Magistrate Court' (Home Office 2008: 3). Within Oxfordshire the court is based within the city, and as yet there are no courts based in the smaller rural towns, which instead must make do with a magistrates' court.

A key service for women and children trying to escape violence comes through the provision of refuge support to supply safe but temporary accommodation (Levison and Harwin 2000). Within Oxfordshire there is provision for 13 beds in Oxford city, 12 beds in North Oxfordshire and a specialist four-bed 'Asian' women's refuge, which can only support single women. The interviewees revealed that the majority of women accommodated within the Oxford refuge come from outside the area, further impacting on this essential provision. The work of Coy *et al.* (2009) highlights the fact that within Britain less than two thirds of all local authorities have refuge spaces, and that there is even less provision for women and children from BME backgrounds. Oxfordshire is lucky that it has at least some refuge provision, although it is woefully insufficient to meet demand.

Partnership-working in a Rural Setting

As discussed earlier within this chapter, multi-agency working is an integral part of community safety delivery (see Hughes 2002). The 1998 Crime and Disorder Act requires local authorities, the police and partner agencies to devise clear and auditable clear strategies for the reduction of crime within local authority areas, including domestic violence. Almost all agencies working within the community, in whatever capacity, have been enrolled into this crime reductive milieu.

The majority of those questioned in Oxfordshire believed that such multi-agency working within the rural areas of the county was markedly better than in the city, for several reasons: violence was more widespread in the city and therefore services were stretched further and staffing was not consistent at the multi-agency meetings and therefore relationships could not be built and maintained. None of those questions was able to provide significant evidence of specific help dedicated to rural areas. Certainly they were not addressing the key issues facing domestic violence victims within rural areas highlighted earlier in the chapter, such as lack of transport or feelings of isolation. One of the main problems highlighted by all of the interviewees was the allocation of funding and service provision for the partnership agencies. Provision might well be limited to a specific geographical location, or overlap a county boundary.

Although tentative, the research highlights that the provision of services based within the Thames Valley in Oxfordshire are not aimed specifically towards rural areas (see McCarry and Williamson 2009) and are focused more within the city of Oxford and the larger towns. Services within the South and West of the area, which are the most rural, are based on outreach work with no direct provision being allocated as yet. It is the belief of the authors that a more in-depth audit of services within rural areas is needed to gain a clearer insight as to how allocation of funding and service provision is made.

Conclusion and Recommendations

The preliminary observations we have made in relation to policing domestic violence in Oxfordshire provides some insight into the factors that shape the experience of violence for rural women. It also offers some direction as to the types of interventions that might be necessary to usefully prevent violence against women in rural areas in the longer term. Ironically, although the police may be among the most frequently contacted of resource agencies, women experiencing domestic violence often report that they are the least helpful. Compared to previous research studies on the police response in the UK (Edwards 1989) it is clear that there have been substantial improvements in recent years, reflected in women's positive accounts of the police (Women's Aid 2007). However, unpredictability in the quality of response remains a problem largely attributed to insufficient training and monitoring. If formal agencies are not consistently helpful, then women have no other choice but to try to cope on their own. It is this isolation, compounded in rural settings, which ensures that violence against women remains an endemic problem.

For rural women, then, there is the additional risk that domestic violence will reoccur. Those that take action to prevent it may face further danger because they are both socially and geographically isolated from services such as transport or healthcare (Mackay 2000). If violence against women in rural areas is to be prevented, the allocation of adequate resources needs to be assured. Without adequate funding and the coordination of essential programmes it is difficult for women living in rural areas to weigh-up their options for escaping violence and abuse. Similarly, to focus resources exclusively in one area, such as concentrating on the policing response at the expense of general advocacy, can result in for rural-based women being further neglected by the criminal justice system.

In tackling such a complex and persistent societal problem as domestic violence we would argue that a number of things have to be addressed. One is the absence of a consistent and coherent voice emerging from within the domestic violence sector. The different perspectives that exist within statutory/voluntary and in different women's organisations makes it hard to develop a unified policy framework; when this relates to something as fragmentary as rural policy the need for a strong, coherent voice is even more pressing.

There are also some policy recommendations that would be effective immediately. Safety planning, allowing women to report violence without risk of danger, has to be prioritised in rural areas and this must take into account work that addresses violent men. Central government itself must invest fully in supporting women and children in rural areas who experience violence in their lives through a strong infrastructure that encourages professionals to work in partnership with those experiencing different forms of violence. This will produce sustainable solutions that build upon individual and community strengths to foster resilience against domestic violence in both the short and long term.

Chapter 15

The Thin Green Line?
Police Perceptions of the Challenges of
Policing Wildlife Crime in Scotland

Nicholas R. Fyfe and Alison D. Reeves

Introduction

Concerns about the environment increasingly intersect with issues of criminal justice (White 2008). At a global level, attention has tended to focus on international regulatory initiatives, such as the Convention on International Trade in Endangered Species (CITES) which was introduced in 1973 and aims to ensure that international trade in specimens of wild animals and plants does not threaten their survival. Interpol (the International Criminal Police Organisation) also has interests in the illegal traffic of species of wild flora and fauna and has established a working group involved in both investigatory and operational activities connected with wildlife crime (White 2008: 197–8).

At a local level, however, the focus is less on illegal trafficking and more on those illegal acts that harm wildlife species. In Sweden, for example, there is particular concern about the poaching of large predators and frustration that very few of the perpetrators of this type of poaching are convicted (Brå 2008). In Australia, it is abalone (sea-snails) theft that commands national attention because global demand means that it is industry worth over AUS$100 million a year (White 2008). In Scotland, the focus of this chapter, there are a range of wildlife crimes that have attracted national political attention as the Minister for Environment observed during a Scottish Parliamentary debate about wildlife crime in October 2007:

> Wildlife crime is an issue which is becoming increasingly significant in this country, not least the persecution of birds of prey … However, other issues such as poaching, the illegal fishing of pearl mussels, hare coursing and the importation of rare and protected species – dead or alive – also need our attention. We are all victims of wildlife crime in that it threatens to diminish the rich natural heritage for which Scotland is rightly world-famous and which is of great importance to our economy. (Scottish Parliament 2007, col 2494)

This Parliamentary debate revealed the increasing anxiety among politicians in Scotland about the scale and significance of wildlife crime. Although it is notoriously difficult to measure precisely the true extent of wildlife crime (because of problems of under-reporting and the hidden nature of any crimes against wildlife, see Conway 1999), claims were made during the debate that 2006 had been the worst year ever for recorded wildlife poisoning incidents and the Scottish Society for the Prevention of Cruelty to Animals (SSPCA) had investigated more than 600 calls about wildlife crime in that year. According to an article in *The Scotsman* newspaper published on the day of the debate, Scotland had witnessed significant increases in the number of offences involving birds and badgers and in cases of cruelty to wild animals. The impact of such crimes, Members of the Scottish Parliament (MSPs) emphasised, was not just on wildlife and the natural environment but also on industry given that more than 250 businesses and more than 3,000 employees are involved in the nature and wildlife tourism industry in Scotland (Scottish Parliament 2007: col. 2526).

The Scottish Government's response to these problems has taken a variety of forms. Since devolution several pieces of legislation have been introduced which strengthen the protection of species and allow for harsher punishment of offenders The Criminal Justice (Scotland) Act 2003 created tougher penalties and custodial sentences for a number of offences against wild birds, animals and habitats and also increased police investigative powers and rights of arrest. The Nature Conservation Act (Scotland) 2004 strengthened legal protection for threatened species and police powers of search, added provisions relating to the protection of wildlife habitat sites and outlawed various pesticides some of which had been used to poison wild animals and raptors, while the Animal Health and Welfare (Scotland) Act 2006 included action on animal fights. In addition to this legislation, the Crown Office and Procurators Fiscal Service (COPFS) has appointed a dedicated team of specialist wildlife prosecutors and each of Scotland's eight police forces have wildlife crime coordinators and officers who, to quote one MSP during the Parliamentary debate, 'bring energy, dedication and expertise to their work' (Scottish Parliament 2007: col. 2494).

Despite these developments, however, concerns remain about the ability of Scotland to tackle the problem of wildlife crime effectively. In relation to policing in particular several MSPs voiced anxieties about the level of priority and resources allocated to this aspect of police work. 'We need more wildlife crime officers in our police forces' one MSP observed, 'but the officers must be properly resourced, valued and supported – a special dedicated force of full-time officers' (Scottish Parliament 2007: col. 2514); and another noted 'that it can be difficult for a hard-pressed police force that is fighting drug crime or street violence in their community to give wildlife crime the priority [we] think is necessary'.

In this chapter, we examine the role of the police in the policing of wildlife crime in Scotland in more detail. Drawing on documentary sources and on interviews with wildlife crime coordinators and wildlife crime officers in each of Scotland's eight police forces, we consider police perceptions of the challenges they confront in tackling wildlife crime. Before addressing these issues, however, it will be useful to place this work in a broader disciplinary context by considering the developing field of 'green criminology' and to consider some of the ambiguity surrounding the term 'wildlife crime'.

'Green Criminology' and the Difficulties of Defining Wildlife Crime

Ten years before the debate in the Scottish Parliament about wildlife crime, the Scottish Government funded a pioneering study of crime in rural Scotland (Anderson 1997). This observed that 'almost by definition the countryside is the venue for much crime against wildlife (e.g. deliberate poisoning of birds of prey, or the theft of rare birds or eggs) as well as offences involving environmental damage and poaching'. 'Green criminology', a term first coined by Lynch (1990), refers to work which focuses on issues of environmental harm, environmental laws and environmental regulation (see Beirne and South 2007; South and Beirne 2006; White 2008). Encouraging a reappraisal of traditional notions of crime, green criminology examines the ways in which individuals, groups, corporations and governments can all generate environmental harm, whether this is in terms of so-called 'primary green crimes' (such as air and water pollution deforestation, and species decline and animal rights) or 'secondary green crimes' (which can include the dumping of hazardous waste or state violence against environmental groups). There is clearly considerable overlap between green or environmental crime and wildlife crime in that both are concerned with ecological harm and with issues of non-sustainability in relation to species that are present today but may not be in the future.

However, defining wildlife crime with any precision is challenging. In the UK, a detailed review of relevant legislation and the legal, criminological and ecological literature carried out by the Department for the Environment, Food and Rural Affairs (DEFRA) as part of the government's plans to establish a National Wildlife Crime Unit concluded that, 'Definitions of what constitutes wildlife crime are problematic'. In relation to legislation, for example, the report noted that 'The current legislative map ... reflects changing cultural, social and environmental priorities, and is both patchy and inconsistent in terms of the levels or types of protection on offer to certain categories of wildlife' (Roberts *et al.* 2001: 11). More generally, the report observed that, 'Any working definition ... needs to be inclusive, so that both the offence-related and the impact-related significance of wildlife crimes – national and international – can be fully addressed' (Roberts *et al.* 2001: 18). The working definition used by DEFRA understood wildlife crime to mean 'those illegal acts which destroy, harm, exploit or threaten wildlife species' (13–14).

Further reflection on the difficulties of defining wildlife crime appeared in the 2004 report of the UK Parliament's Environmental Audit Committee inquiry into wildlife crime. The Committee voiced concerned that the absence of an accepted definition 'has had a direct and negative impact on the public's perception of wildlife crime' (Environmental Audit Committee 2004: paragraph 6). They noted, for example, that the public would probably agree that some activities (such as badger baiting and the stealing of eggs) were crimes but few would describe their own actions (such as the destruction of a rare and protected plant during a visit to a Site of Special Scientific Interest) as criminal. Similarly, a property developer who disturbs a bat colony while clearing a site might not view what they are doing as criminal even though this is an offence under UK law. Moreover, the Committee felt strongly that the absence of any agreed formal, informal, legislative or other definition of the term wildlife crime seriously affects the work of those charged with protecting wildlife. 'It is unacceptable', they concluded, 'that those entrusted with the enforcement of our current legislation do not have a clear and agreed definition of the crime they are to police. Without an agreed definition of wildlife crime, which is shared and acted upon by all those who work in the wildlife arena, we believe it is impossible for any real headway to be made in the fight to reduce the number of such crimes' (Environmental Audit Committee 2004: paragraph 8). Like DEFRA, the Environmental Audit Committee resorted to a working definition of wildlife crime as 'any action which contravenes current legislation which governs the protection of flora and fauna' and that can result from 'ignorance, neglect or a deliberate act'.

The Structure of Policing Wildlife Crime in Scotland

From this brief review of the definition and character of wildlife crime, it should be clear that it poses significant policing challenges. As White observes in relation to environmental law enforcement:

> Crimes may have local, regional and global dimensions. They may be difficult to detect ... They may demand intensive cross-jurisdictional negotiation ... Some crime may be highly organised and involve criminal syndicates, such as illegal fishing. Others may include a wide range of criminal actors, ranging from the individual collector of endangered species to the systematic disposal of toxic waste via third parties. (White 2008: 197)

Against this background, there is an immediate question concerning who should police wildlife crime. Should it be left to specialist agencies with powers to investigate arrest and prosecute or should the police service have direct involvement in this field? Within Scotland, the view has been expressed by Her Majesty's Inspectorate of Constabulary for Scotland (HMICS) and the Crown Office and Procurators Fiscal Service (COPFS) and endorsed by government, that wildlife crime should be treated 'like any other crime' and therefore the police should have a key, although not exclusive role, in its prevention and investigation (Scottish Government 2008). The structure for delivering this policing function has both national and local dimensions. At a national level, the Association of Chief Police

Officers in Scotland (ACPOS) has set out its intentions with regard to wildlife crime in its 'Strategy for dealing with wildlife crime 2006/8' (ACPOS 2006). In addition there is a Scottish Wildlife Crime Tactical and Coordinating Group (SWCTCG) which produces a control strategy based on the strategic assessment of wildlife crime. These national level arrangements came under critical scrutiny in 2008, however, as a result of an inspection of the arrangements for the prevention, investigation and prosecution of wildlife crime carried out by HMICS and COPFS (The Scottish Government 2008). This highlighted that only a minority of Scottish police forces were familiar with the ACPOS strategy and that the SWCTCG control strategy was poorly resourced and therefore had little impact on individual forces. The report was unequivocal in concluding that there is 'a systemic breakdown in the national police management of wildlife crime in Scotland'.

At a local level, Scotland's decentralised structure of eight police forces means there is significant variation in how wildlife crime policing is delivered by individual forces. Although all forces have a senior lead officer for wildlife crime and all have wildlife crime coordinators, these are only full-time positions in four forces and, of these, only two have appointed regular officers (of Sergeant and Inspector rank) with civilians fulfilling the role in the remainder. All forces also have wildlife crime officers (WCOs) for whom wildlife crime is an additional responsibility to their other duties as local police officers. The way in which WCOS are deployed, however, varied between forces. Some WCOs are used as a force resource allowing officers to travel across divisional boundaries to deal with wildlife crime, while other WCOs are restricted their own divisions (Scottish Government 2008). Other local differences between WCOs included the selection process, with some forces allowing this to happen via self-nomination and others requiring an interview involving the force coordinator.

In the following two sections we draw on interviews carried out with both wildlife crime coordinators and WCOs in each of Scotland's eight police forces in order to examine their perceptions of some of the internal, police organisational challenges and external, community-based challenges associated with tackling wildlife crime.

'Bunny-Hugging Type Stuff': The Internal Challenges of Policing Wildlife Crime

In his review of policing and environmental law-enforcement, White (2008) notes that senior police managers face difficult decisions in deciding how best to allocate their limited resources. 'Public opinion, media and political attention and internal policing dynamics', he observes, 'will all affect if, why and how, specific types of environmental crime are addressed. How environmental issues are perceived within a police service will inevitably have an impact on organisational priorities' (White 2008: 203). In this section, we draw on interviews with wildlife crime coordinators and WCOs to show how the perception of wildlife crime within police organisations affects the priority given to investigation and enforcement in this area.

Limited Resources and the Marginalisation of Wildlife Crime within Police Organisations

During its inquiry into wildlife crime in England and Wales, the Environmental Audit Committee heard evidence from Chief Constables about their reluctance to divert resources to address an issue which they felt was not a government priority:

> Chief Constables undoubtedly have sufficient resources to deal with wildlife crime should we decide that such matters be priorities. However, we receive no messages from government indicating that these matters should have resources directed towards them. Few chief Constables are therefore prepared to dedicate resources towards areas they are not asked to concentrate on. (Environmental Audit Committee 2004)

The experience in Scotland is not dissimilar. The combination of part-time wildlife crime coordinators in four forces and WCOs having to juggle their work on wildlife crime with their other responsibilities creates a situation where many officers feel their impact and effectiveness is very limited:

> But what you have to remember is that a Wildlife Crime Officer like myself, I am a full-time employed Police Officer, so you know yesterday I was dealing with crime, as in crime that we recognise, not wildlife crime. But last night I was up until one o'clock this morning with daily talks and things to do with wildlife crime … So if it wasn't for the officers working on their days off and in their own time then it, you know we just couldn't cover the ground. (WCO: interview)

This quotation is interesting in the remark about the distinction between wildlife crime and 'crime that we would recognise'. It suggests that the limited resources allocated to wildlife crime policing may be linked to a wider issue concerning the marginalisation of wildlife crime within police organisations. Such marginalisation partly reflects the organisational priorities of police forces. As one wildlife crime coordinator explained:

> Because the house break-ins and thefts … are classed as high priority, wildlife is not, so I can't go away and be proactive in dealing with wildlife crime because there's more issues which are of a higher priority. So in the structure of a force … wildlife crime would be close to the bottom. I would probably say it's the same in all forces: wildlife crime is a low priority. Not sure the

[Scottish] Executive would like to hear me saying that but it is. (Wildlife crime coordinator: interview)

This was a view echoed by officers in other forces. One officer spoke of the 'resistance' by divisional officers to addressing wildlife crime issues because they are 'well down the pecking order'. Reinforcing this perceived organisational marginalisation of wildlife crime, however, was what some wildlife and environmental crime coordinators and WCOs viewed as a trivialisation of wildlife crime as not being 'real police work' or part of 'real policing'. Ethnographic studies of front-line policing routinely drawn attention to the 'cop cultural' distinctions between 'real police work' focused on crime fighting and the pursuit of 'good class villains' and other types of police work which are much more to do with order maintenance and service delivery. Such cultural assumptions clearly have a relevance to the policing of wildlife crime as well. As one coordinator put it: 'To the cops on the ground, some of the older ones still think it's sort of bunny-hugging type stuff, because it's new to them and it's not something they grew up with in their service'. Another example, illustrating differences of opinion within a police force about whether an example of wildlife crime is in fact a police matter, is highlighted in this observation made by a WCO:

> When I sit in on meetings and operations that go through the major incidents over the past twenty four hours, … wildlife crime is not taken so seriously because of stabbings etc and okay I can understand it. But all I'm asking for is these guys to be allowed to deal with crime, that's what they are paid to do, just happens to be wildlife crime but that is still not seen as real policing. I had an incident where there was a seal killed and the Inspector on duty said it's nothing to do with the police and I had a real battle with him. 'It's a police matter', 'No it's not', 'Yes, it is'. This is an Inspector. So if an Inspector's saying that, what chances have officers got to go and deal with it. (WCO: interview)

Countering Marginalisation: The Importance of the National Intelligence Model

Such exchanges clearly indicate the continuing challenges facing coordinators and WCOs in terms of securing a place for wildlife crime on their force agenda. Significantly, however, both wildlife crime coordinators and officers spoke of a variety of strategies that can help counter the marginalisation of wildlife crime within their forces. One is ensuring clear support for wildlife crime among the most senior officers within a force. Another strategy is to raise the profile of wildlife crime at the grass-roots level of the organisation to such an extent that it is recognised as core police business. An example of the latter, as one WCO explained, simply means changing the way wildlife incidents are handled when reported to police:

Before, through no fault of the operator who was taking the call, [wildlife crime] would just disappear into the fog. If somebody's phoned in and said, 'Well I have seen a trap, I think it's illegal', … [the operator would say] 'Well, hang on you need to inform the SSPCA [Scottish Society for the Prevention of Cruelty to Animals] and give them a ring'. But we've raised the profile, we've raised the awareness, so if somebody was to phone in, say 'we've seen two guys going into a wood' that would automatically be an incident created … And that way the whole profile has been raised and people are genuinely happy after the response that they get now. (WCO: Interview)

Closely connected to this profile raising activity was a concern among wildlife crime coordinators and WCOs to highlight the links between wildlife crime and other forms of criminality. This emerged as a significant theme in the Environmental Audit Committee's inquiry into wildlife crime in England and Wales where the Committee observed that the 'refusal to accept wildlife crime as an issue deserving of committed police resources [is] especially short-sighted given the many links made between wildlife crime and serious and organised crime' (Environmental Audit Committee 2004: 11). In our interviews this too emerged as a key concern of wildlife crime coordinators and WCOs who cited many examples of cases where those involved in wildlife crime were typically also involved in other, often, organised, criminal activity:

Locally, hare coursing, has a significant impact. Usually committed by people who are on the fringes of organised crime in that field anyway, I don't like stereotyping, we're not supposed to do that but it's a fact. It's very rarely, committed by somebody who doesn't have a record. (Wildlife crime coordinator: interview)

Such links between wildlife crime and other forms of criminality are significant not just at a cop-cultural level, but also in terms of the wider organisational perception of wildlife crime. Crucial in this regard is the process of embedding wildlife crime within the National Intelligence Model (NIM). The NIM provides a common basis for gathering and analysing information and using this to adopt a problem-solving approach to reducing crime and disorder (Herdale and Stelfox 2008: 173). Central to the NIM is the notion of decision-making based on the production of intelligence products, such as subject profiles. In situations where there is intelligence that an individual is involved in wildlife crime as well as other forms of criminality, this can be brought together via the NIM in the form of a subject profile and form the basis of a business case to use police resources to target that individual. As one coordinator explained:

Through the national intelligence model we have a control strategy which basically sets out your priorities for the year. And I'm afraid wildlife crime is way down the list, you know, which is a problem. But it's a problem which we have to overcome and I see us doing that by increasing intelligence. I see it as not trying to combat wildlife crime on its own … I know there is a case in X at the moment, a guy who is actively, quite regularly, he's coursing with dogs. And he's taking down, or the dogs are taking down roe deer. He then goes and he slits their throat. Now that's a bad crime. However, people would laugh at me if I went to try to get a team together to combat that. But the guy's a druggie. And he's also run about with a firearm. So if you put the three things together, and there's a lot of information coming in regarding the drugs, there's a chance I might get something done about that. But the point I'm making is, to go on the wildlife angle alone, which I've learned very quickly, you're wasting your time. You've got to have it combined with something else. (Wildlife crime coordinator: Interview)

The significance of using the NIM as a vehicle for levering resources within police forces for tackling wildlife crime should not be underestimated. As another coordinator explained:

If wildlife crime doesn't feature in the NIM, it's never going to feature in the Force's priorities … So somebody like me that's sitting here as X can see what an operational PC probably can't, in that if you get wildlife crime to feature on the NIM radar it's going to get resourced … So I've been starting to try and introduce NIM terminology into the management of wildlife crime and it's having an impact to the extent that we've got a full-time Wildlife crime officer now we didn't have a year ago … My previous two predecessors had both submitted a business case each of which got the knock-back. I built on their business cases and I built on the reasons why they got the knock-back and I realised that the main thing was they weren't weaving the application into the NIM. (Wildlife crime coordinator: Interview)

Using the NIM therefore provides a mechanism for levering resources by linking wildlife crime to other forms of criminality.

'We're Not Lone Rangers': The External Challenges of Policing Wildlife Crime

In this section we turn from internal, police-centred, challenges of policing wildlife crime effectively, to consider some of the external, community-based challenges. In particular, we highlight police concerns about the contested character of wildlife crime and the nature of liaison with other agencies.

Contesting Criminality: Local Cultures and the Social Construction of Wildlife Crime

Earlier in this chapter we highlighted the debate surrounding the definition of wildlife crime and how the public might view something like badger baiting as criminal but not necessarily think that disturbing a bat colony is also a crime. Such confusion does not simply reflect imperfect knowledge of relevant legislation but points to a wider and more fundamental question concerning the social construction of environmental problems. 'When it comes to environmental harm', White observes, 'what actually gets criminalised by and large reflects the efforts of claims-makers to make an issue of the harm in question' (White 2008: 46). Moreover, understanding the socially constructed and therefore contested criminality of environmental harm draws in 'Moral positions, ethical principles, traditional understandings and common sense knowledge' (White 2008: 53). These are not just matters of abstract debate. For police officers, these issues surrounding the social construction of environmental crime within communities and the way in which the criminality of particular activities and practices are often contested because they are rooted in local cultural or historical traditions, are things they routinely confront. As one wildlife crime coordinator explained:

> Where it's in the communities, it's what's always been done ... I was actually speaking at a school in [location deleted] and this was a secondary school telling them about offences against badgers and this 14 year old girl said, 'my Dad and my Granddad were out catching badgers last night'. 'They're breaking the law'. 'No, they're not. My Dad says it's okay to do that' ... It's just what they do. (Wildlife crime coordinator: Interview)

The involvement of the girl's father and grandfather in this example highlights the importance of tradition in how attitudes towards wildlife in communities typically have strong historical roots and the way in which established practices may predate the subsequent criminalisation of these activities. A second example of the contested character of wildlife crime relates to the re-introduction of sea eagles to Scotland. These birds became extinct in the British Isles in the early part of the twentieth century but, beginning in 1975, there has been a long term project to re-introduce them by importing eaglets under special licence from Norway. The latest phase of this project, which involves the Royal Society for the Protection of Birds (RSPB), Scottish Natural Heritage (SNH) and the Forestry Commission, began in June 2008. Support for this initiative, however, has varied and one wildlife crime coordinator explained how the re-establishment of sea eagles in northern Scotland is often problematic for some local people. Evidence of this concern made the national (UK) news in September 2008 with claims by Scottish farmers in the north-west Highlands that these birds had killed more than 200 lambs in the previous 12 months. Speaking to the BBC, the chair of a local crofting association explained that this year has been particularly bad:

The crofters know how many lambs they put out after lambing season and one woman lost fifty percent of her animals. She actually saw a sea eagle lifting a lamb from her field and flying off with it … We have been keeping lambs here for generations and if this is not sorted out, this could be the end. (BBC News 2008)

Such claims are contested by conservation organisations who argue that although sea eagles will take lambs they are normally already dead or weak and it is unlikely that the birds are alone responsible for the loss of lambs in the area. Other birds of prey too are often blamed for losses as well. As one wildlife crime coordinator explained, 'if you take the eagle owl, I mean [you get] horrendous stories in the press: they're not going to be hovering above like primary schools waiting to take kids at playtime. And oh, Mrs Smith's cats disappeared, you know, they'll get the blame for it'.

In terms of addressing this challenge of contested criminality, wildlife crime coordinators and officers typically view themselves as in the frontline in terms of reshaping attitudes towards wildlife particularly in terms of what might be termed 'early intervention' work they do in schools to educate children about the nature of wildlife crime and its consequences before they become inculcated with the norms of older generations. This point has a wider resonance with respect to the social control of environmental harm. While the police clearly have a reactive role in relation to environmental harm *vis-à-vis* the investigation and prosecution of wildlife crime, they clearly have also developed a pro-active role in respect to education and development of ecological citizenship.

Partnership Working, Prosecution and Wildlife Crime

In policing wildlife crime, the police don't just interact with local communities but also work in partnership with a range of other organisations such as the Royal Society for the Protection of Birds (RSPB), the Scottish Society for the Prevention of Cruelty to Animals (SSPCA), and Scottish Natural Heritage (SNH). The police bring particular skills and powers to such partnership-working based on their experience of conducting investigations and capacity to enter premises, seize materials and arrest and detain suspects. Previous research (Conway 1999) has noted the good relationships between the police and organisations like RSPB, SSPCA and SNH and this was echoed in our interviews with wildlife crime coordinators and WCOs. Nevertheless, it was also clear that effective partnerships require trust and that one of the challenges facing WCOs is to build social capital in the form of networks of relationships based on norms of trust and reciprocity. As a WCO explained:

Okay so I need to have a firm local knowledge of my target species, … I need to know where the Red Squirrels are likely to be, I need to know where the badger sets are, and this is where getting involved with the other groups was extremely hard. I have to say, they did not trust me to start with and I can fully appreciate and understand why. 'Who is this policeman coming along here? What does he do in his day off? We're trying to protect the world's global population of badgers and here he is wanting to know where the sets are, can we trust them?'. That's, that's a healthy thing, even, towards distrust of myself, but as the years have gone on and I've broken down the barriers, we have a really good working relationship. (WCO interview)

In addition to working with particular organisations, the police also work with key individuals in rural areas, such as farmers and gamekeepers. Previous research in Scotland highlighted how the perception of the police and gamekeepers 'being seen as too friendly' (Conway 1999: 26) could hamper relationships with organisations such as RSPB who might be concerned about sharing information with the police about, for example, the location of a sensitive nest site. However, in our interviews, the police were more concerned that in the past they had perceived people like farmers and gamekeepers as 'the enemy' in the sense that they might be responsible for committing wildlife crime and that the challenge facing the police now was to 're-think' who the enemy is by treating farmers and gamekeepers as potential allies rather than potential offenders:

Now the so-called enemy, being stereotypical and blinkered, you could say well it must be the crofters and the farmers and the gamekeepers and we said 'No, it can't be, these people are the eyes and ears of the country, these are the people who are looking after the countryside', and they don't always get it right, the sea eagle persecution is one small example but in the main they do get it right, ... They look after the local environment, they kind of nurture it, and they understand the life cycle of things, so it works well together so why don't we get in touch with these people? And instead of making enemies of them, the 'them' and 'us' scenario ... I was going round and structuring talks to the college for gamekeepers and now on a regular basis I go and I speak to these young, up and coming gamekeepers, and befriend them and we have a working relationship now. They have a rural watch as a result of all this and we now have twenty four people dotted around ... who are land managers, crofters, estate workers, hoteliers, and they're absolutely fantastic, they're dedicated. (WCO interview)

A significant aspect of partnership working concerned the relationships with personnel from other parts of the criminal justice system and, in particular, those in COPFS are responsible for prosecuting wildlife crime. Two issues emerged. The first focused on concerns about the knowledge and skills of those in the prosecution service to handle wildlife crime cases. As one WCO observed:

You spend hours, days, maybe hundreds of man hours, getting a conviction, going for a case to submit to the Crown Prosecution Service or to the Procurator Fiscal, depending on what side of the border you're on, and it can all fall flat and the person can get off with it because maybe the judicial system isn't really quite clued up on wildlife crime and maybe the judge or the magistrate really doesn't think that it's very significant to lock this person up because he had a couple of eggs in his pocket ... So we have to convince them and it can be extremely frustrating when a case is thrown out as not in the public interest to prosecute because maybe it's the first offence and it's the only time that this criminal had ever been caught dealing with a wildlife crime. (WCO interview)

However, the joint HMICS and COPFS thematic inspection of wildlife crime suggested the system is working better, concluding that, 'overall COPFS had introduced a sound system for managing wildlife crime, particularly by establishing a network of specialist wildlife prosecutors who should prosecute these cases' (Scottish Government 2008). The second area of concern mentioned by police officers related to the impact of sentencing and what they perceived as the relatively lenient sentences for some types of wildlife crime. Whether this perception of leniency with respect to the prosecution of wildlife crime is accurate is open to debate. The joint thematic inspection carried out by HMICS and COPFS reviewed some 80 wildlife crime cases from January 2006 to November 2007 and of these 56 proceeded to court and 44 resulted in a conviction. In 10 of the cases which resulted in conviction, the court ordered no financial or custodial penalty and instead required alternatives such as community service. Where there was a conviction and financial penalty, the average penalty for each accused was almost £500 (compared with an average of £300 for each accused in all cases irrespective of subject matter). More recently, in 2008, the owner of a Scottish grouse moor had his £107,000 farming subsidy cut by the Scottish Government because of suspicions that pesticides found on his land were being used against birds of prey. This was a record civil penalty imposed under European legislation which makes the protection of wildlife a condition of the subsidy (*The Guardian* 2008).

Conclusions

'For the police', White (2008: 197, 203) observes, 'dealing with environmental harm is basically dealing with the unknown ... Environmental law enforcement is a relatively new area of police work ... and is at a stage where perhaps more questions are being asked than answers can be provided'. In Scotland, however, as this chapter has illustrated, the police have developed a wide range of experience in dealing with wildlife crime against a background of considerable legislative activity aimed at protecting Scotland's natural heritage.

Nevertheless, the police still face significant challenges in providing an effective response to wildlife crime. Some of these challenges reflect the location of wildlife crime relative to other police organisational priorities but other challenges are more community-based and focus attention on locally contested definitions of 'wildlife crime' and 'environmental harm' when legislation is seen as conflicting with traditional rural practices. This chapter has provided some insights into these issues and there are similarities with findings from other studies, such as that carried out by the Swedish National Council for Crime Prevention on the poaching of large predators, such as wolves and golden eagles (Brå 2008). In Sweden, as in Scotland, definitions of wildlife crime are contested, with hunters and livestock owners perceiving the spread of large predators as a problem and typically receiving strong support for predator poaching from members of their local communities despite it being unlawful.

More generally, however, the paucity of research evidence in this field means there is clearly scope for further study. Such research will have an important role to play in informing practitioners and policy makers about the challenges of policing wildlife crime, highlighting both organisational issues (such the resources committed to wildlife crime and the need for partnership working) and cultural issues (such as the perceived legitimacy of laws and regulations relating to wildlife). In addition, however, more research will also contribute to developing a green criminological agenda which, to date, has tended to be dominated by studies of state and corporate environmental crime. Indeed, according to White (2008) one of the key tasks of green criminology must be to study the nature of environmental regulatory mechanisms and the social control of environmental harm. Investigating the policing of wildlife crime by both 'the police' and other agencies provides one way of contributing this wider theoretical agenda but also addresses issues of pressing policy concern. As one of the contributors to the wildlife crime debate in the Scottish Parliament (discussed in the introduction to this chapter) observed:

> Wildlife crime is a serious matter … We should care about it because cruelty to animals is totally unacceptable and has direct links with other forms of human violence and degrading behaviour … Scotland's wildlife protection laws may be among the best in Europe but our record on the effective prevention, investigation and prosecution of wildlife crime could be better. (Scottish Parliament 2007: column 2513).

Acknowledgements

An earlier version of this chapter was presented at the Stockholm Criminology Symposium in June 2008 and we are grateful for participants for their comments. We would also like to thank the Carnegie Trust for the Universities of Scotland for funding the field work on which this chapter is based and for the cooperation of Scotland police forces.

Chapter 16

Policing Poaching and Protecting Pachyderms: Lessons Learned from Africa's Elephants

A.M. Lemieux

Introduction

The poaching of wild animals might be considered a crime by society against nature. The term poaching describes a number of illegal actions that directly harm animals and threaten the sustainability of their populations; including killing or trapping animals (see also Fyfe and Reeves, this volume). It is a socially contested activity that reflects a complex set of interactions that transgress physical, legal and cultural boundaries within and between states. Although the act of poaching occurs in specific rural or wilderness sites, the transportation and sale of poached goods reflects a network of illegal activities that extends globally and connects a wide range of people and places. The policing of poaching must therefore operate across a range of scales and, consequently, faces many challenges.

In light of these issues, this chapter presents a typology for understanding how the demand for wildlife products is created and sustained on a local and international level. It will use poaching of the African Elephant as an example of how this typology can be applied when analysing poaching problems and devising prevention measures.

Poaching and the African Elephant

The African Elephant is found in 37 sub-Saharan countries that are known as the 'Elephant Range States' (Blanc *et al.* 2007). Concerns for their survival led to the implementation of monitoring programmes that revealed a steep decline in numbers, from 1.3 million in 1979 to fewer than 600,000 a decade later (Figure 16.1). International poaching for ivory was thought to be driving this collapse and in 1989 the Convention on International Trade in Endangered Species of Wild Fauna or Flora (CITES) recognised that African Elephants were 'threatened by extinction and in dire need of help from the international community'[1] (CITES

---

1   Indicated by the inclusion of African Elephants in Appendix 1 of the CITES register (Messer 2000).

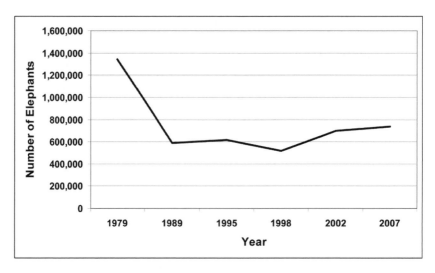

Figure 16.1   Africa's elephant population (1979–2007)

Sources: van Aarde and Jackson 2006; Said *et al*. 1995; Barnes *et al*. 1999; Blanc *et al*. 2003; Blanc *et al.* 2007

2008). This categorisation attempted to diminish ivory markets by banning the international trade of elephant products between member nations of CITES.

CITES is an international agreement between the governments of 173 member countries that seeks to protect endangered species of wildlife by banning or limiting the international trading of specific animals and animal products. In the years following the ban on ivory trading, the African elephant population grew to just over 730,000 animals.[2] Yet these figures mask considerable differences between the elephant populations of different regions (Figure 16.2) and poaching remains problematic in many central and western African states (Stiles 2004; Blanc *et al*. 2007). Thirteen out of 20 nations in Central and West Africa have significant problems with elephant poaching whereas elephants are only threatened in four of the 16 nations in the South and East regions of Africa. According to analyses of elephant carcasses by CITES in 2003/4, 80% of elephants died as a result of poaching in central African countries, compared to 7% and 24% of all recorded deaths in South and East Africa respectively (CITES 2004). Of the 17 nations with acute poaching problems, 13 are in Central and West Africa (Barnes *et al*. 2007).

While the CITES convention provides a framework to guide the application of relevant state laws, CITES regulations are enforced nationally. Because of this the international ban on ivory is only as strong as the political will and resources allocated to its implementation. It is, therefore, important to understand how and

---

2   These estimates include both Forest and Savannah Elephant.

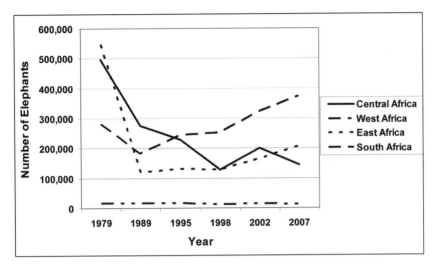

Figure 16.2   Elephant populations by region (1979–2007)

*Source*s: van Aarde and Jackson 2006; Said *et al.* 1995; Barnes *et al.* 1999; Blanc *et al.* 2003; Blanc *et al.* 2007

why poaching occurs and what can be done at the local, state and international level to police it. With this in mind, the following section identifies the roles of different actors within poaching networks.

Poachers – Traders – Transporters: A Useful Typology

Poachers, traders and transporters of wildlife all contribute to the illicit killing, sale and movement of animals. By placing these actors into a typology, it is easier to understand how demand for wildlife products is created and how poaching can be prevented by targeting each group simultaneously or independently. In essence, poachers obtain wildlife products illegally, traders buy and sell these products, and transporters help move the illicit items between destinations. This typology is meant to help academics and local authorities categorise poaching problems and develop better, more effective prevention measures. It can easily be applied to any type of poaching.

Poachers

The three tasks poachers must accomplish to supply ivory markets are: (1) kill an elephant illegally, (2) remove its tusks, and (3) store them safely. Killing an elephant requires that hunters enter a public or private reserve unnoticed and use their weapon of choice to take the animal's life. Once dead, removing the ivory can be done with any hard, sharp instrument such as a machete, axe or hatchet. With the ivory in-hand a poacher's last objective is to exit the setting. This means he must carry the tusks out of the preserve on foot or with a vehicle. Poachers then find a safe place to store their gains until they are ready to be sold. Criminal opportunities for these actions arise when access to parks and weapons is unmonitored.

This chain of events must happen before anyone can trade poached ivory. If a poacher is unable to kill an elephant, removing its tusks is no longer an option. Likewise, anti-poaching efforts may force a hunter to dump or hide his gains in a national park before he is able to store them safely. In either case, poached ivory does not enter the market and cannot be sold.

Traders

Traders buy and sell ivory in various places around the globe. In general, the first trader to enter the elephant poaching equation is the poacher himself. Unless an animal has been poached for personal consumption, hunters sell ivory to any number of people. It is likely that these tusks will be resold more than once and eventually be carved into some type of decorative or functional piece. There are numerous markets for ivory around the world. Of main concern are the unregulated ivory markets in Africa (Table 16.1) because they are not controlled or monitored by local governments. In countries with regulated markets, traders must register their inventory and provide documentation that proves the ivory they sell is not from poached animals. Their unregulated counterparts do not have these requirements and provide poachers the opportunity to introduce illicit ivory.

Transporters

Transporters move illicit ivory within nations and across international borders. There is no volume requirement for transporters and they should not be pigeonholed as a group who only moves large shipments. For example, a tourist who buys ivory chopsticks in Asia and brings them home to North America is a transporter. This group builds markets by increasing the distance over which African elephant products can be used and sold. They have various modus operandi that depend on the level of scrutiny their shipment will face in-transit. This means avoiding detection may involve false documentation, painting or staining ivory, not declaring purchases at customs, exploiting diplomatic cover, or bribes.

Table 16.1   The role of poachers, traders, and transporters in Africa's elephant range states

Region	Country	Poaching?[1]	Ivory Market?[2]	Transit Route?[2]
Central	Cameroon	Yes	Unregulated (large)	Yes
	CAR	Yes	Unregulated (medium)	No
	Chad	Yes	Unregulated (small)	Yes
	Congo	Yes	Unregulated (medium)	No
	DRC	Yes	Unregulated (large)	No
	Eq. Guinea	No	Unknown	Unknown
	Gabon	Yes	Unregulated (small)	No
East	Eritrea	No	Unknown	Unknown
	Ethiopia	Yes	Unregulated (large)	Yes
	Kenya	No	None	Yes
	Rwanda	No	Unregulated (small)	Yes
	Somalia	No	Unknown	Unknown
	Sudan	No	Unregulated (medium)	Yes
	Tanzania	No	None	Yes
	Uganda	No	Unregulated (very small)	Yes
South	Angola	No	Unregulated (small)	Yes
	Botswana	No	Regulated (very small)	No
	Malawi	Yes	Regulated (small)	Yes
	Mozambique	No	Unregulated (large)	Yes
	Namibia	No	Regulated (very small)	No
	South Africa	No	Regulated (large)	No
	Swaziland	No	Regulated (very small)	No
	Zambia	Yes	Regulated (very small)	Yes

	Zimbabwe	Yes	Regulated (very large)	No
West	Benin	No	Unregulated (very small)	No
	Burkina Faso	Yes	Unregulated (very small)	No
	Cote d'Ivoire	Yes	Unregulated (large)	No
	Ghana	No	Unregulated (small)	No
	Guinea	No	Unknown	No
	Guinea Bissau	No	Unknown	Unknown
	Liberia	No	Unregulated (very small)	No
	Mali	No	Unregulated (small)	Yes
	Niger	Yes	Unregulated (very small)	No
	Nigeria	Yes	Unregulated (very large)	Yes
	Senegal	Yes	Unregulated (medium)	Yes
	Sierra Leone	Yes	Unregulated (very small)	No
	Togo	Yes	Unregulated (small)	No

[1]*Source*: Blanc *et al*. 2007

[2]*Source*: Milliken 2004

These seemingly simple interactions are actually quite complex and involve numerous feedback mechanisms that influence levels of poaching. For example, if a trader can sell large quantities of product he will increase poaching incentives. However, if that same trader depends on a transporter to move his product into the international market and the transporter is apprehended, poaching incentives decrease as supply accumulates. Further, these networks extend across international boundaries, making it difficult for one state to disrupt them unilaterally. Indeed, the balance of these activities varies considerably between states (Table 16.1) so that many countries with large ivory markets do not have elephants and very often ivory is sold far from its point of origin. This has implications for the type and scale of policing needed to disrupt these networks. Given that 23 countries in Table 16.1 have unregulated markets, greater efforts should be made by these states to control the trading of ivory. Likewise, countries with transit routes are increasing poaching incentives: closing these routes while targeting poachers would likely result in faster and more sustained declines in illegal hunting. The following sections examine ways in which illicit poaching, trading and transporting activities can be policed.

Targeting Poachers

Patrolling Elephant Range

Illegal hunting is the first crime in a series of events that determine the international demand for ivory and so the prevention of illegal hunting is perhaps the obvious way to limit ivory markets. But as there are more than 3.3 million km² of elephant range in Africa (Barnes *et al.* 2007), often in remote and wilderness areas, and it requires enormous financial and human effort to police it constantly. For example, the cost of patrolling the 14,000 km² Central Luangwa Valley of Zambia, one of the poorest nations in the world, was estimated at $82 per km² per annum in the 1990s (Jachmann and Billiouw 1997). Yet, as this area demonstrates, locally targeted initiatives that are properly resourced and funded, including personnel costs, can be effective in reducing hunting in particular places. The next sections describe tactics and technology that can be deployed by local wildlife officers.

Low-Cost Solutions

Low-cost techniques are capable of limiting access to elephants and decreasing poaching incentives. Any low-cost tactic should try to make it harder for poachers to find and kill elephants.

Situational methods (Cornish and Clarke 2003) include closing roads or monitoring traffic. In Central Africa, 'poaching is exacerbated by new roads for logging operations and mineral and oil extraction, which provide both access to deep forest and routes for the transport of ivory and meat' (Barnes *et al.* 2007). Road closures prevent hunters from using vehicles to access the remote areas where elephants live and make carrying their tusks to safety more difficult. As poachers on foot move more slowly than those in a vehicle, this method may increase chances of their apprehension. If closures are not possible, simply monitoring the traffic on logging roads may deter potential poachers. The main objective of these techniques is to limit contact between hunters and elephants.

A more controversial option for poaching prevention is shoot-to-kill legislation. In short, these laws give wildlife officers the right to shoot and kill any person who is illegally hunting elephants. This is considered a low-cost method because it does not require any additional weapons or officers. Even if a park is managed by five part-time officers, the efficiency of those officers is greatly increased by the ability to draw their weapons on those who break the law. The legislation is largely symbolic and is aimed at the rationality of hunters. In Zimbabwe, the number of elephants poached declined from 4,000 in 1989 to 46 in 1992 when the shoot to kill policy was enacted (Kelso 1993).

Such policies raise serious concerns. Poachers may more likely to kill guards, witnesses or potential witnesses to their activities, prompting an escalation in the use of arms by guards and poachers with lethal consequences. Clearly, legal, ethical and moral questions also arise about the use of summary justice to kill poachers on sight/site. This type of legislation is best viewed as a symbolic, last-resort solution. A more desirable approach is to control weapons and ammunition. By limiting the sale of elephant guns and large calibre ammunition, local authorities could make poaching much more difficult at a relatively low cost. This may not be an effective option in regions with high gun ownership and civil conflict, but could work in stable nations where dealers are supplying hunters with weapons specific to elephant poaching.

While these methods are reactive, more proactive methods of community policing may yield better results in the long term. In short, agencies should be encouraged to engage and work with local communities to prevent poaching. Communities near poaching hotspots may have information about the modus operandi of hunters and opening dialogs with these groups may help wildlife officials understand what is driving the poaching and how it can be prevented. More broadly, working with rather than against local communities will increase understanding of the problems associated with poaching and may lead to longer-term improvements. Such actions should be part of wider sustainable development initiatives that improve wealth and reduce the need to hunt illegally for economic subsistence.

High-Cost Solutions

A range of technologies can be employed to prevent or deter poaching. However, they are expensive to install, limiting their availability to many African States, and their effectiveness may be limited by terrain and other factors. The most obvious example is aeroplanes which can be used to perform population counts of the local elephant population, look for poached carcasses and track illegal hunters. They are, of course, very costly to run and are not effective in areas with thick canopy covers, especially in Central Africa (Walsh and White 1999).

Surveillance systems can also provide situational defences against poaching. These include sensors that detect metal in weapons or microphones that can triangulate the location of gunshots. Both have the potential to act as deterrents and to increase the response times of security agencies when crimes have been committed but are locality specific. They need to be placed in areas known to be vulnerable to poaching or trafficking, such as known trails in the case of metal detectors. Thus, like many target-hardening approaches to crime prevention, poaching may be displaced to other places where these systems are not used.

Controlling Traders and Transporters

Whereas efforts to reduce hunting are focused on specific localities within countries, market reduction and other demand focused approaches require international cooperation. An excellent example of this is the CITES ban on ivory trading that aimed to decrease the worldwide demand for ivory. The ban *has* helped Africa's elephant population rebound in some countries although those in Central and West

Africa are still losing animals through poaching (Figure 16.2).[3] This is because the CITES ban did not provide universal funding for anti-poaching efforts and it appears that elephants only benefited from the ban in countries with adequate conservation budgets. This highlights the need to compliment demand reduction strategies with supply-focused methods. Further, the trade ban will only remain effective if it is supported by effective custom controls that prevent both tourists and large-scale smugglers taking ivory between countries.

Tourists are major traders and transporters of ivory (Courouble *et al.* 2003; Martin and Milliken 2005; Milliken 2004; Milliken *et al.* 2006; Sheperd 2002; Wu and Phipps 2002) and a range of carrot and stick approaches can be used to reduce their involvement. On the one hand, efforts could be made to educate and inform them about the (un)ethical, (il)legal and (un)environmental aspects of the ivory trade through leaflets, displays at airports and so on in an effort to persuade them not to buy ivory souvenirs. Where this fails and tourists, through ignorance or willing, endeavour to take ivory goods home, then rigorous customs enforcement aims to prevent physically their removal and to prosecute those aiming to do so. CITES recommends the use of long wave ultra-violet light and magnifying glasses to distinguish ivory from common substitutes (Eapinoza and Mona 1999).

In 1997, CITES adopted a resolution to create a 'comprehensive international monitoring system ... to track illegal trade in elephant products' (Milliken *et al.* 2007). This system is known as the Elephant Trade Information System (ETIS) and has been compiling country level data on ivory seizures since 1999. The analysis of ETIS seizure data for patterns in trade routes would allow a profile of transporters and their packages to be developed, aiding interception. This may increase the efficiency of screening procedures and better focus resource allocation. Similarly, DNA technology can be used to detect the origin of illegal shipments of ivory. Using small samples, a group at the University of Washington is able to reference tusk DNA to samples taken from elephant dung across the continent. It recently identified Zambian elephants as the source for the largest seizure recorded since the 1989 ban (Wasser *et al.* 2007). In general, the more officers know about poachers and traders in their locale the better they will understand the movements of transporters.

Conclusion

Poaching is the result of poachers, traders and transporters creating criminal markets that demand wildlife products. These groups work together to optimise financial gain from the illegal hunting and trapping of animals. Preventing poaching requires local and international communities to understand their role in

---

3   For detailed analysis of how elephant populations have changed on a country level, see Lemieux and Clarke (2009).

the wildlife trade. Some places may have all three groups working side by side; others will only have one or two.

Limiting the number of criminal opportunities available to poachers is the most straightforward way to prevent poaching. This can be done by limiting access to areas where wildlife live, controlling weapons, increasing police patrols, and the use of technology to enhance the monitoring capabilities of wildlife authorities. Poaching can also be prevented by reducing the demand for wildlife products. These measures should target traders and transporters. Disrupting markets and increased screening of imports and exports are examples of this.

African elephant poaching was used as an example of how to apply the typology presented. It was found that poachers, traders and transporters of ivory are not evenly distributed across the continent or world. This leads to the conclusion that local authorities will best prevent elephant poaching by understanding what role their country plays in supplying or creating the world's ivory markets. Not every country in Africa needs a large budget for anti-poaching patrols. In fact, just 17 of the 37 range states in Africa have elephant populations threatened by poachers. Most nations facilitate poaching by creating demand for the illicit product and are not directly linked to the supply. This means international efforts need to address the problems of individual countries. The poacher-trader-transporter typology presented here can be used as a framework for resource allocation and legislative reform. It is meant to help law enforcement officers, academics, and government officials describe local problems and devise appropriate solutions.

Chapter 17

Policing Agricultural Crime

Joseph F. Donnermeyer, Elaine M. Barclay and Daniel P. Mears

Introduction

The idea that farms are crime-free has never been true. However, the industrialisation of farming in advanced capitalist countries has made agricultural operations and their vast array of expensive machinery, equipment and supplies bigger targets for crime than ever before (Barclay and Donnermeyer 2007).

There are two broad sets or categories of crime that can occur on farms. The first includes so-called 'ordinary crimes': offences which have been part of the farm scene for many decades, such as the theft of livestock, machinery and farm supplies, vandalism, rubbish dumping, and damage from trespassers and hunters (Barclay and Donnermeyer 2007; Mears *et al.* 2007a, 2007b, 2007c). A second set of crimes has emerged in more recent times that seems 'extraordinary' for their potential impacts, even though their occurrence may be many times less frequent than ordinary crimes. One new threat is organised drug production, such as marijuana and methamphetamines, that is conducted on remote and hidden areas of farms without knowledge of the operator. This is especially the case for clandestine methamphetamine labs, whose environmental impacts create severe hazards for farm workers and animals as well as the environment (Donnermeyer and Tunnell 2007; Weisheit 2008). Terrorism is a second type of extraordinary farm crime. Sometimes referred to as bioterrorism or agro terrorism, the sabotage of food production systems has become a recent concern to many nations (Bryne 2007; Moats 2007). Both types of extraordinary crimes are worth considering, even though their occurrence may be rare, because they illustrate the geo-political contexts of agriculture and its integration into global systems of trade.

What We Know About Agricultural Crime

Systematic research on farm crime is very sparse. Essentially, there have been two waves of farm crime studies. The first is a small cluster of victimisation studies conducted 15 plus years ago in the US states of Alabama (Dunkelberger *et al.* 1992), Arkansas (Farmer and Voth 1989), Kentucky (Peale 1990), Mississippi (Deeds *et al.* 1992), Ohio (Donnermeyer 1987), Tennessee (Cleland 1990) and West Virginia (Bean and Lawrence 1978). The second wave is more international, including research reports from Australia (Anderson and McCall 2005; Barclay *et al.* 2001, 2007; Barclay and Donnermeyer 2002; Barclay and Donnermeyer 2007; Carcach 2002; Donnermeyer and Barclay 2005), the state of California in the USA (Chalfin *et al.* 2007; Mears *et al.* 2007c), England (Sugden 1999) and Scotland (George Street Research 1999). Their findings are remarkably similar, both in terms of victimisation rates and the geographic factors associated with crime, suggesting a common ecology for farm crime in these countries (Barclay *et al.* 2001; Barclay and Donnermeyer 2007).

Based on this work it has been estimated that between 10% to 26% of agricultural operators have experienced machinery and equipment theft, 13% to 24% vandalism and 6% to 12% burglary. Livestock theft ranged from less than 1% to as much as 13% annually, depending on the location. When comparisons are possible with suburban and urban locations, rates of victimisation against farms are often much higher (Barclay and Donnermeyer 2007).

The studies also found a specific ecology to a farm's vulnerability. Larger-sized farms, operations near well-travelled state/provincial and national highways, and those closer to large towns or cities were more likely to experience a crime, especially theft (Barclay and Donnermeyer 2007; Mears *et al.* 2007c). Agricultural operations with scattered plots of land and with storage of machinery, equipment and supplies at isolated locations also experienced higher rates of theft. Operations farther away from urban areas but near a public road experienced a greater amount of trespassing (often by hunters), vandalism, and rubbish dumping (or fly-tipping). Several studies found burglary to the farm homestead to be extremely low, but the rate of break and enter of farm buildings to be quite high. The study by Barclay and Donnermeyer (2002) also found in the Australian state of New South Wales that farms with more vegetation and a hilly terrain experienced higher rates of victimisation, especially for trespassing, illegal hunting and livestock theft.

These results indicate two interrelated findings that hold true for many agricultural regions: crime varies inversely with guardianship (i.e. informal policing by the operator and family members, farm workers, and neighbours) and directly with accessibility. Consider, the study from the Midwestern US state of Ohio (Donnermeyer 1987). Most commercial farms in Ohio and other Midwestern states are several hundred to a thousand acres in size. In a typical Midwestern farm, the homestead is the place where the farm family lives, and near most homesteads is a central area where many of the day-to-day tasks necessary to run an agricultural operation occur. Many, but not all, of the barns and other farm buildings are within walking distance of the house. There is usually a family member or farm worker (or even the family dog) at or near the home, hence, break-ins and thefts are much less frequent because guardianship is a natural consequence of daily work activities. However, most of these operations also have farm buildings and storage facilities located out of sight of the homestead due to terrain, distance, and trees and bushes. It is at those buildings that the overwhelming share of breaking and entry crimes occur. Guardianship is much lower, and usually, there is a rudimentary dirt or gravel/stone lane from a public road leading to the storage facilities, enhancing accessibility for both the owners, and as well, the thieves. In drier climates than the Midwest, such as Australia, where agricultural operations are much larger in size, buildings may be even more spread out and farm work may not be as close to the homestead, but the dual principles of guardianship and accessibility still apply. Agricultural crime more likely occurs when guardianship is low and accessibility is high.

As both sources of evidence demonstrate, agricultural security is indeed a costly and serious problem. Using our own research on agricultural crime from New South Wales (Barclay *et al.* 2001; Barclay and Donnermeyer 2002; Barclay and Donnermeyer 2007; Barclay and Donnermeyer 2009; Donnermeyer and Barclay 2005) and central California (Chalfin *et al.* 2007; Mears *et al.* 2007a, 2007b, 2007c), this chapter goes on to discuss issues related to the policing of agricultural crime.

Policing Agricultural Crime

Even though the specific circumstances or risk factors associated with agricultural crime appear to be similar, it should not be presumed that issues related to the policing of farm crime are equally similar. Issues of policing, for example, in New South Wales and central California will be dependent on the organisation of law enforcement at these places. In New South Wales, there is a single police agency, with officers assigned to rural communities on a rotating basis (Scott and Jobes 2007). As an earlier study on rural crime in a small town of New South Wales (O'Connor and Gray 1989) noted, residents anxiously awaited the re-assignment of officers in terms of their background. Did the officer grow up on a farm or in a rural community? Would the officer understand agriculture and the norms and values that make smaller Australian communities different from big cities, like Sydney?

In the USA, police are organised very differently. The USA has a so-called 'federated' system, with national, state and local levels of governance. It is essentially law enforcement in local municipalities and open-country and farm areas who would be concerned with the investigation of 'ordinary' farm crime, and who would sponsor prevention programs and enforcement initiatives.

However, such extraordinary agricultural crime as organised drug production and agro-terrorism can include state-level police and various enforcement agencies at the national level. Adding to the complexity of the US system is the fact that responsibility for various police functions for most agricultural areas is in the hands of sheriffs (Weisheit *et al.* 2006). Sheriffs are elected, usually to a four-year term. Even though the candidates must be trained and certified as law enforcement officers (a requirement stipulated by state legislatures only over the past several decades), the office of sheriff remains a political position. In US counties where agriculture is economically important, it likely carries more political clout. Hence, the political economy of a county may determine how much attention is paid to the safety and security needs of farmers.

Australia

Agriculture utilises almost 60% of Australia's total land area of 7.7 million km², and the 155,000 agricultural businesses are mainly engaged in livestock production (beef cattle, dairy cattle and sheep) and/or growing grain (Australian Bureau of Statistics 2008). The theft of livestock is the most significant rural crime in Australia where thieves are encouraged by the isolation of farms, the unlikelihood of detection and high financial rewards. The financial cost and the loss of valuable breeding stock and blood lines are significant for farmers. There is also the threat to biosecurity as thieves rarely consider Australia's strict quarantine measures. Livestock thefts range from a single animal for food to thefts of large numbers of stock committed by highly skilled, well-equipped, and well-organised professional thieves. Other major farm crimes include the theft of fuel, trespassing, illegal hunting, and the dumping of rubbish. Further, due to the isolation of rural areas, cannabis can be cultivated on properties where there are large areas that are rarely entered by the owners (Barclay *et al.* 2001; Barclay and Donnermeyer 2002).

In Queensland, New South Wales, Victoria, Tasmania, and the Australian Capital Territory, there are specialised police squads or rural crime investigators (RCIs). In the other states, local police investigate farm-related crime as part of their normal duties. The primary function of RCI's is the investigation of stock related crime. This involves attending farms to investigate suspected livestock offences, conducting musters to identify stolen livestock, and monitoring livestock movements at abattoirs, saleyards and roadways to detect possible thefts. Other duties include ensuring relevant statutes are enforced and participating in various types of operations such as static transport interceptions, targeting drug and property offences, and assisting in searches for drug crops or missing persons (Queensland Police 2008). Biosecurity has become a major focus of RCI teams in recent years with concerns about possible disease outbreaks that would devastate Australia's livestock industries. Recently an outbreak of Equine Influenza highlighted the role of police in limiting the spread of the disease which was largely due to human error (Callinan 2008).

RCIs are equipped for extensive travel across remote areas with laptop computers, video and digital cameras, satellite phones, UHF radios and camping equipment. They also have horses, trail bikes, stock trucks and 4WD vehicles. RCI's work closely with their counterparts across state borders, local communities, farmer organisations and various government agencies (Queensland Police 2008).

A study of 36 RCI's in New South Wales (Donnermeyer and Barclay 2005) found the most common obstacles for effective policing were the lack of time and resources to adequately deal with farm crime. It takes time to patrol roads, stop trucks and check them over, complete the necessary paper work, visit sales and abattoirs, talk to people, and gather information. Many rural officers reported the area they had to cover was so large and diverse that it was impossible to patrol it all effectively. In one instance it took two officers five hours to attend and investigate a single farm crime scene. Police response is also influenced by the relative importance of farm crime in the police agenda. Scarce resources of staff, time, vehicles and equipment to pursue investigations into farm crime have forced police to weigh the importance of these types of offences against the needs of all crime in rural towns.

Although livestock theft caused major financial losses as well as the loss of breeding stock, few offences were reported to police. Victims believed police could or would not do anything about it, and felt that the police (especially those from the city) have little knowledge or understanding of agricultural industries. Officers believed the reluctance to report crimes or pass them information about suspicious activities or people within the district was part of a rural culture where there is an inherent tolerance of stock theft and a tendency for people to keep to themselves and not report crimes (Donnermeyer and Barclay 2005).

Police exercised wide discretion in their response to calls for service from farmers. Some first considered the management practices of the farmer before concluding whether or not a report was credible. Others were suspicious that some farmers reported thefts for the purpose of tax evasion. Police were frustrated by poor management practices and the lack of precautionary behaviours among farmers. Farmers are mostly nonchalant about security, rarely locking gates, sheds and fuel tanks or locating stock yards away from the road. The lack of identification on livestock, farm produce, machinery and equipment and poor record keeping of stock numbers and stock movements were described by officers as the greatest barrier for controlling and preventing farm crime. In any investigation, police need to know exactly what is missing (Donnermeyer and Barclay 2005).

In court, farmers must be able to provide proof of ownership. One officer maintained that gathering sufficient evidence to prove a case of stock theft beyond reasonable doubt is one of the greatest challenges for police. Offenders easily escape conviction by pleading ignorance of the theft. Although livestock theft carries a maximum of 14 years' imprisonment, usually offenders get off with a bond. Some officers believed that some magistrates, who come from the city, have little understanding of agricultural crimes and the culture within rural communities. As the courts are too lenient and the penalties are so weak and infrequent, farmers are discouraged from helping or advising police (Donnermeyer and Barclay 2005).

As the costs of farm crime in Australia have increased, rural officers attempt proactive solutions to the problem. One initiative is Rural Watch, which is an extension and adaptation of the neighbourhood watch schemes found in many cities. Success was dependent upon the interest and enthusiasm of the members and the social value of the group and cohesiveness of the community. Groups worked better when farms were relatively close together. In outback regions, distance between properties rendered rural watch ineffectual. However, farmers used UHF radio to maintain a 'virtual' rural watch in the district (Donnermeyer and Barclay 2005).

One solution to the problem of stock identification and traceability is the National Livestock Identification Scheme (NLIS). The system was introduced to improve trace-back and monitoring systems for stock diseases and chemical residues and to allow Australian producers to compete on the international market where similar systems are being considered or have already been adopted. Using machine readable radio frequency identification devices on animals, this centrally-stored electronic history of an individual animal enables rapid and accurate traceability of livestock or carcasses in the event of a disease outbreak or chemical residue incident and can also aid crime prevention by providing a means of tracing stolen stock (Animal Health Australia 2008).

All in all, there are two layers or levels of reality in the policing of farm crime in Australia. The first relates to the challenges of enforcement in wide-open areas where guardianship on the part of the farm operator and the police is difficult. The second relates to local cultures, and the kinds of social class distinctions that can be found in all rural communities. Officers' decisions as to how they use their time in relation to agricultural crime are conditioned by both local geography and a local cultural-political context.

California

As discussed earlier, a number of state-specific studies in the USA were undertaken in the 1970s and 1980s, but interest in the topic waned and it remains the case that little is known about the prevalence of agricultural crime nationally, the causes of such crime, or how best to prevent or reduce it. One recent and notable exception stands out. During the past decade, the US Department of Justice helped to fund an agricultural crime initiative in California's Central Valley, which was conducted by Mears and associates (Chalfin *et al*. 2007; Mears *et al*. 2007c). Originating out of the Tulare County District Attorney's office, the initiative, called the Agricultural Crime, Technology, Information, and Operations Network (ACTION) project, consists of five activities aimed at preventing and reducing agricultural crime: (1) collecting and analysing agricultural crime data; (2) encouraging and enabling information-sharing among law enforcement agencies and prosecutors within and across counties; (3) educating the public and farmers about agricultural crime and how to combat it; (4) marking equipment with owner applied numbers (OANs); and (5) promoting aggressive law enforcement and prosecution (Mears *et al*. 2007a). The results of the federally-funded study of ACTION are described in several reports and articles (Chalfin *et al*. 2007; Mears *et al*. 2007a, 2007b, 2007c). We focus here on several key themes related to policing and agriculture.

First, the study reinforced the findings of other studies in showing that agricultural crime, including the theft of livestock, commodities, equipment, and chemicals, not only is relatively common but also rarely results in any kind of response from farmers, law enforcement agencies, or the courts. That is true even for more serious types of theft, such as tractor or chemical theft. But it is especially true for lesser offences, including tool theft and crimes that the public might consider nuisance offences, such as dumping trash on the edge of a farmer's property. Hence, any proactive attempt to address agricultural crime holds great potential for generating substantial reductions and diffusing awareness of the seriousness of farm crime and how to prevent it to other regions of the country and beyond the USA to other countries.

Second, the California study confirmed what other research typically finds: there are certain predictable features of agricultural crime that in turn point to ways in which prevention efforts should be targeted. To illustrate, it is evident that certain crops are at greater risk of theft when they are in season or when various market forces drive prices up. Such information can be used, but is frequently not, to guide law enforcement efforts. For example, during the ACTION evaluation, it appeared that when the prices of avocados rose, so too did their theft. Such theft is rendered easier in many instances because farmers frequently grow avocados on hillsides where the crops can readily be seen. In this case, however, 'Guac Cops' who specialised in agricultural crimes, helped farmers develop approaches to protecting their crops (such as putting fences around crops, patrolling their property, and using guard dogs) and they also undertook several sting operations aimed at helping prosecutors convict individuals who participated in high-volume theft (Brown 2004).

Third, the fact that crime prevention efforts likely must take account of unique features of some crimes means that a single approach cannot work equally well with all types of theft. For example, reducing the exposure of various targets, whether they are crops, livestock, tools, or the like, can help reduce the likelihood of victimisation. In that sense, there is a universal step, exposure reduction or 'target hardening', that can be taken. But what that step will look like in practice will vary greatly depending on the type of crime. For example, although it may be possible to lock pesticides in a building, it may not be feasible to store all large equipment in a building at a centralised location; in the latter instance, parking equipment in areas hidden from public roads is the more viable option.

Fourth, it is not only the characteristics of the specific crimes but also such factors as the seasonality of certain crops and commodities that dictate the nature and intensity of an intervention. Using the avocado theft example, in cases where fences cannot be cost-effectively installed, it makes little sense to patrol avocado crops year-round. Rather, the more cost-efficient strategy lies in intensifying patrol efforts at times of year when various crops or commodities are particularly vulnerable to illegal harvesting.

Fifth, farmers are vulnerable to victimisation from multiple sources, and so an effective response to reducing agricultural crime requires a multi-pronged strategy. At a typical farm, theft can be undertaken by many individuals, including employees at the farm or neighbouring farms, individuals who happen to be driving by, and those who are involved in organised crime efforts. Juxtaposed against the many potential offenders are the innumerable opportunities for theft, especially in contexts where the ability to monitor all aspects of an operation frequently are limited, and the limited targeting of agricultural crime by law enforcement or the courts. This constellation of factors gives rise to a need for agricultural crime prevention efforts that adopt a wide range of strategies, including ones that address farmers' behaviours, that reduce opportunities for offenders to vandalise or steal from farms, and that increase the actual or perceived penalties associated with agricultural crime.

The ACTION initiative also considered the need for public understanding of the problem. The focus on the public stemmed in part from a recognition that public understanding and pressure sometimes are required to generate sufficient interest by law enforcement agencies and the courts to tackle specific types of crimes. Fighting gang problems or increases in homicides constitute easy sells in most jurisdictions. By contrast, agricultural crime is little understood and in most jurisdictions would seem relatively trivial. That said, in regions where farming constitutes a core economic activity, members of the public may well be more willing to support or call for efforts to address it.

ACTION's focus on law enforcement and prosecutorial efforts stemmed from insights, too, about the critical role that such efforts can play in promoting more active responses from farmers and in elevating agricultural crime as an issue warranting public attention. If law enforcement agencies do not respond in a proactive way to calls from farmers about various types of agricultural crimes, it reinforces a message to farmers that, indeed, such crimes are a cost of doing business and that nothing can be done about it. Further, if officers aggressively respond to such farmers' calls but prosecutors do little to help build agricultural crime cases, it would be reasonable, if unfortunate, for officers to cease paying much attention to agricultural crime. In short, any effective attempt to enlist law enforcement agencies in efforts to combat agricultural crime requires, at the same time, enlisting the assistance of farmers, prosecutors and the public.

The challenges of creating collaborative efforts between these different groups are difficult to understate. As but one example, in the hierarchy of units or divisions within U.S. law enforcement agencies and prosecutors' offices, agricultural crime lies near the bottom, if it is recognised at all. In this regard, perhaps the cornerstone of ACTION's efforts is the emphasis it placed on generating a collaborative ethos and set of strategies both within and across the different counties that comprise California's central valley. The ACTION staff worked carefully to promote buy-in from the different groups within and across the counties, including farmers, law enforcement agencies, and prosecutors, and to facilitate communication among these groups.

Not least, ACTION prioritised the generation and analysis of agricultural crime data. Many jurisdictions' law enforcement agencies and court systems do not systematically code for specific types of agricultural crimes, and so it is frequently not possible to generate real-time accounts of such (reported) crime. ACTION developed a database that directly confronted this problem by collecting law enforcement and prosecutor data in such a way as to be able to readily compile county-level descriptions of reported agricultural crime, arrests, and prosecutions. It also included address-level information about specific crime incidents, thus enabling ACTION staff, in conjunction with the county law enforcement agencies, to produce crime maps that depicted areas where certain types of agricultural crime were concentrated or where there appeared to be shifts in such crime from one area to another. Such information was useful in, among other things, enabling law enforcement officers to target their sting operations and their outreach efforts.

Finally, ACTION facilitated efforts to mark farmers' equipment with owner applied numbers (OANs). Typically, when equipment is stolen, there is little ability to recover it or to know who owns it. By contrast, marked equipment can be easier to recover and to return to owners. In addition, it can serve as a potential deterrent to offenders because, if caught with the equipment, law enforcement agents can more easily determine if it was stolen.

Given the multi-faceted nature of the ACTION initiative, it was and remains difficult to undertake a rigorous impact evaluation linking particular activities (e.g. OAN marking or prosecutions) to specific outcomes (e.g. increased crime prevention steps among farmers, reduced agricultural crime of various types). Even so, it was evident, for example, that ACTION was successful in promoting more within- and between-county collaborations within and between law enforcement agencies and prosecutors' offices; that farmers increased their crime protection measures; and that prosecutions of agricultural crime increased.

Conclusion

Both the focused and highly coordinated approach of California's ACTION program and the various initiatives in a number of states in Australia share one common trait: a recognition that the vulnerability of farms is greatly influenced by ecological factors of space and time. There is no other type of business that shares these same features. Remoteness, isolation, type of terrain, density of cover or vegetation, and low population density creates points of high vulnerability on many farms. The increased cost of equipment and supplies and the seasonal nature of most farming activities enhance the likelihood of crime against agricultural operations as never before. Farm property of various kinds are more desirable targets, and the utilisation of equipment and supplies during times of peak work activity on a farm presents problems of storage (hence security).

Added into the mix are factors of politics and community values and norms which place pressure on local farmers to not report crime (unless covered by insurance) and affect the decision-making/discretionary behaviour of the police. All of these factors point to the way that agricultural crime can only be understood in the context of broad social structural and economic change that has transformed the context within which food is produced and security threats to agriculture have grown.

One research related recommendation is to begin longitudinal studies of crime against farmers, utilising a similar methodology as the National Crime Victimisation Survey of the US Department of Justice. Households in the NCS participate for several years, with interviews conducted periodically in order to frame the reference period for recording of crime events. Once benchmark or base data is established, participating farms can be used to field test prevention technologies and actions, with the possibility of sub-dividing the sample into experimental and control groups. Already in Australia, the installation of cameras at remote entrances to paddocks is underway, creating the possibility for an analysis of their deterrent effect. GIS/GPS technology can be used to identify possible 'hot spots' related to the ecology of farms and their probability of becoming victimised by various crimes.

The recommendations above rely on micro-theories of place, such as routine activities theory and situational crime prevention theory (Donnermeyer 2007a). Certainly, much more research at this level is necessary and welcomed.

However, there is much more to do. In a recent article, Donnermeyer and DeKeseredy (2008) call for a new rural critical criminology in the same vein as the pioneering work of Taylor and Young (1973). To quote: 'It is time to think critically about rural crime and theoretical work on this topic needs consistently and explicitly to address the unsettling truth that one key determinant of crime is structured social inequality' (Donnermeyer and DeKeseredy 2007: 17). This is no truer for agricultural crime than it is for violence in rural communities, or any other form of crime as it occurs in the rural context. Unfortunately, there is not a single farm crime study that takes a critical approach, yet, there is research that hints at ways that the discourse on the policing of agricultural crime can be expanded to consider both a broader structural and a more critical point of view.

The first study comes from Northern Ireland, which geographically is not far from the research conducted in England (Sudgen 1999), and in Scotland (George Street Research 1999), but by context, is very different. Armstrong (2005) examined farm victimisation for the county of Fermanagh, which borders the Republic of Ireland. Through key informant interviews and police statistics, Armstrong (2005) noted that most farmers believed illegal dumping or fly-tipping was the single largest problem, often with 'weekend trippers' from the towns using the countryside to dispose of their garbage. Similar to the research from Australia, the theft of 'quads' (four-wheel drive vehicles with cross-country capabilities) from farms was the most frequently reported form of theft. So far, his findings look much the same as other farm victimisation surveys, but what was different about Armstrong's investigation was the cross-border context in which these incidents occured. Farm machinery stolen from Northern Ireland were frequently taken to the Republic of Ireland through Fermanagh. Armstrong (2005) describes the border of Fermanagh as 133 miles long and possessing a 'maze of inter connected roads' (p. 40). Further, many police stations have closed in the rural areas of Fermanagh subsequent to the 1998 peace accord signed by the Protestant and Catholic factions in Northern Ireland. Previously, these stations functioned to help maintain border security, but also represented a very strong police presence that deterred much crime. With the peace accords, police resources shifted to cities and towns. As a result, crime against rural residents of Fermanagh, especially farmers, has grown quite rapidly, along with a rather dramatic increase of concerns and fear about crime.

A study of crime in Madagascar (Fahcamps and Moser 2003) found an inverse relationship between crime and population density, which is the opposite of criminological assumptions for advanced capitalist societies. The more isolated parts of the country had the highest rates of crime, homicide, rape, burglary and cattle-theft. The incidence of reported cattle-theft was related directly to the size of the operation. The authors attribute a great deal of it to 'banditry,' which are rural crime gangs called 'dahalos'. As well, in the southern sections of Madagascar, cattle thefts reflect ethnic customs related to rites of passage to manhood and eligibility for marriage. Without a strong police presence, similar to what Armstrong (2005) found for Northern Ireland, and aided by isolation, low population density, and poverty, rustling is likely to remain high, and on occasion, other violent crimes take place in association with incidents of stock theft. Similarly, Smith *et al.* (2001) found that among an ethnically diverse set of east African pastoralists in northern Kenya and southern Ethiopia, many of the semi-nomadic groups measure wealth, bravery and desirability as a marriage partner by herd size, hence, 'livestock raiding' or theft by young males is very commonplace. Embedded in centuries of cultural traditions, livestock theft increases the chances of inter-ethnic conflict during times of drought, as it becomes the most viable method for replenishing herds.

Finally, Walters (2004, 2006) speaks about agricultural crime in a different way, namely how crime issues are related to biotechnology and genetically modified foods (see also Henry, this volume). In the earlier publication, Walters discusses the results of the New Zealand Royal Commission's examination of the risks associated with GMO's. They warn of the ways private, corporate ownership of the intellectual rights associated with biotechnology, and the influence of globalisation on all forms of agricultural production has encroached into nearly every part of the world. Walters (2006) stresses the need for a critical look at new varieties of agricultural crime. In this case, agricultural crime and issues of policing become associated with economic exploitation of farm labour and food producers throughout the world, especially in less advanced capitalist countries. This form of 'state-corporate' crime (Walters 2006: 26) considers a 'green criminology that integrates the concepts of power, governance, globalisation and capital with a model of environmental justice' (Walters 2006: 39).

What do these highly diverse pieces of scholarship, as well as the more mundane studies discussed earlier, all touching on aspects of crime, policing and agriculture, have in common? The answer is that future research and theorising about agricultural crime must consider larger, social structural and cultural factors. Within and between the rural sectors of societies, food producers and the police occupy the same positions of privilege and/or disadvantage as other members of society. They are embedded in local, regional, national and even international networks that define differential relationships of power and inequality, and access to capital, technology, and markets. Agriculturalists and local law enforcement are affected by political, ethnic/race and other forms of rivalry and even conflict within their own countries and of cross-border conflicts as well. Often, agriculturalists and rural police go about their work and way of life on the geographical, social, and economic peripheries of their societies. All of the research cited in this chapter, plus speculations about the influence of drug production and terrorism, illustrates how an occupation and way of life which many people, including criminologists, still regard as both crime free and relatively unaffected by broader, global forces related to economic, social and cultural change, is in fact highly vulnerable to a variety of crimes, each of which can only be understood in terms of their place in larger social systems.

Chapter 18

Policing the Producer: The Bio-Politics of Farm Production in New Zealand's Productivist Landscape

Matthew Henry

Introduction

In a recent review of rural policing Buttle (2006: 45) has highlighted the paucity of criminological research into rural crime and policing, and argued that from those studies which do exist, 'it is apparent that rural crime is a rich and complex social phenomena and there are few assumptions that can be made in any given rural context.' While Buttle successfully questions taken-for-granted assumptions as to the geography of rurality and rural crime he leaves related questions as to the plural character of wider networks of rural policing largely unexamined. Notwithstanding a belief that, 'It would be useful to catalogue all the agencies (governmental/nongovernmental) that deal with numerous social problems often linked to crime' (Buttle 2006: 11) the review is framed by the assumption that rural policing is primarily concerned with the maintenance of social order, defined explicitly through the law and/or informally through sets of tacit knowledges and carried out by police who are regarded as the state's primary agents of ordering.

Crawford (2003) offers us a reconceptualisation of the police which replaces the notion of them as the monopolistic keepers of social order with the more fragmented terrain of plural policing. In words that prefigure Buttle (2006) in the contemporary context, 'The police are now part of a varied assortment of organisations with policing functions and a diffuse array of policing processes' (Crawford 2003: 136). Consequently following Crawford the activity of policing is not necessarily defined by its institutionalisation in 'the Police'. If 'the Police' do not define policing, then what else might define it? Here Crawford (2003) identifies three key dimensions that define policing in ways that take it beyond the police: it involves intentional action undertaken by individuals or organisations conscious of their exercise of power; it is directed towards the maintenance of defined norms; and it involves action in both the present and the future. Defined in these terms we can clearly argue that policing is as much a function of education, medicine or welfare agencies as it is of the police. Policing defined in these terms represents a mobile rationality of governance dispersed through myriad state agencies, using multiple technologies and unevenly brought to bear upon emerging problematisations (Rose 1999). This chapter argues that in New Zealand maintaining agricultural production emerged during the 19th century as a field of problematisation, and that the state progressively developed a distinctive regime of agricultural policing that came to encompass New Zealand's farmers and which was primarily concerned with maintaining and improving production rather than maintaining social order.

To situate this argument the chapter fashions a narrative tying together two significant moments in the institutionalisation of agricultural policing: first, the establishment of New Zealand's Department of Agriculture during the late 19th century; and second the recent formation of a new state owned enterprise (SOE) AsureQuality Ltd in 2007. In tracing the emergence of these specific agents of control the chapter suggests that they are framed by a wider logic of biopolitical concern, the recognition of which helps us to understand why agricultural production came to be constituted and embraced as a field of intervention.

Rural Security

Just prior to New Zealand's 2008 general election Federated Farmers of New Zealand (FFNZ), the main farmer lobby group, released its election manifesto. Entitled 'New Zealand's Economic Backbone' the manifesto stressed the continuing importance of agriculture to New Zealand's economy and the challenges to farming posed by issues such as the Resource Management Act (RMA), climate change, biosecurity, the regulation of water resources, taxation and local government funding (Federated Farmers of New Zealand 2008). One specifically identified issue was that of security with FFNZ arguing that because of the isolation of many rural communities they were suffering poorer access to emergency and security services and in some instances a conflict of norms that was eroding community confidence in the police and other emergency services. Concerns which reflected a long history of unevenness (Butterworth 2005; Dunstall 1999; Hill 1995) as well as recent cases, most notably that of Iraena Asher which had drawn attention to the uneven rural reach of the police and emergency services (Corboy *et al.* 2005).

The manifesto also highlighted security concerns which did not directly relate to maintaining social order in rural communities but rather to the management of New Zealand's agricultural industries and landscapes. Thus FFNZ maintained that as long as agriculture provided the basis for New Zealand's economy, 'biosecurity must retain a strong focus on protection of land based primary production' (Federated Farmers of New Zealand 2008: 12); and that in relation to pest control it, strongly supported 'efforts to control and if possible eliminate plant and animal pests that adversely impact on agricultural production'(Federated Farmers of New Zealand 2008: 26). FFNZ's comments sketch the existence of an assemblage of discourses and practice that frames rural ordering in ways that are not easily reducible to a singular attention to the police as the only agents of security. FFNZ pointed to the existence of regimes of governance constituted beyond what we might consider the state's traditional organs of order and whose focus lies not on the maintenance of social order *per se* but rather the governance of agricultural production and productivist landscapes.

Productivism in this context refers to a 'commitment to an intensive, industrially driven and expansionist agriculture with state support and based primarily on output and increased productivity' (Lowe *et al*. 1993: 221). Whilst the understanding and critique of productivism as articulated by Lowe *et al*. emerged in an European context, as Jay (2007) suggests it also describes very well the forces that shaped New Zealand agriculture over the past century. Forces that were given clear expression in New Zealand at the end of Second World War when the Director-General of Agriculture, E.J. Fawcett (1945: 2) reported to Parliament that, 'the Dominion has a clear objective – the development of production to the maximum.' A sentiment which reflected the legacy of development over the preceding century of European settlement and helped cement productivism as the master narrative of New Zealand's agricultural policy following the Second World War.

Producing Productivist Landscapes

The productivist ethos which developed to encompass New Zealand's agricultural producers originated in the dramatic nineteenth century transformation of New Zealand's landscape. This transformation involved the large scale alienation of land from Maori (New Zealand's indigenous peoples) and the transfer of that land into the hands of a colonising population whose primary social, political and economic relationships lay stretched across the globe (see Belich 1996; King 2003). This transfer enabled the widespread clearance of the country's forests and the establishment of grasslands that by 1881 covered 1.4 million hectares and by 1925 covered 6.6 million hectares (almost a quarter of New Zealand's total land area) (Brooking and Pawson 2007). The growth of these grasslands was necessary to the establishment of a pastoral economy which initially relied on the production of wool, but with the advent of refrigerated shipping in 1882 refocused on the production of meat and dairy products for the UK market. For almost 80 years this combination of products and market constituted the bulk of and primary destination for New Zealand's exports (Evans 1969). Whilst the relative importance of these specific products declined in the late 20th century the agricultural sector has remained an enormously significant contributor to New Zealand's material economy and cultural fabric (Williscroft 2007).

The establishment of a pastoral economy resulted in significant planned and unplanned flows of new biota into New Zealand. The dramatic and often unexpected effects of these new biota on the New Zealand landscape was famously described by Gruthie-Smith (1921/1999) and Clark (1949). It is in response to the effects of these flows that the first eradication programmes concerned with 'pests' and disease were developed during the latter half of the 19th century. An early and enduring concern with sheep farmers was the sheep disease scabies which had arrived in Nelson from Merinos imported from Australia in the 1840s (Wolfe 2006). In response the 1844 Sheep Ordinance established a system of inspection and movement controls to try and prevent the spread of the disease (Carter and MacGibbon 2003). However, it was the 1871 Diseased Cattle Act (DCA 1871) which first institutionalised the biosecurity technique of quarantine that would provide the basis for a sanitary regime that continues to the present. Alongside quarantine a key feature of the DCA 1871 was the establishment of a system of regional inspectors with extensive, coercive powers to require farmers to destroy infected stock. It was this aspect of the Act which led the Bill's mover George Waterhouse (NZPD 1871) to reflect on the tension in the legislation between making the system of control collectively effective and trying to ensure that in doing so it inflicted as little individual hardship as possible. What began to emerge then, albeit in an ad hoc fashion was a system of policing whose primary purpose was the protection and improvement of agricultural production and profitability rather the maintenance of social order.

The broader logic of this emerging web of agricultural control can be given some shape if we step back and consider Foucault's (1991) account of disciplinary and biopolitical modalities of power. The development of a web of agencies collectively concerned with security and policing follows from what Foucault (1991) argues was the consolidation of modalities of power which were less interested in the maintenance of power and more concerned with social and economic improvement. From the 17th century onwards Foucault identified in western Europe the emergence of an ensemble of knowledges and techniques founded on the individualisation and socio-spatial ordering of bodies, what he famously termed 'disciplinary power' (Foucault 1977). However, alongside the spread of disciplinary power which focused on the individual body, Foucault suggested that what was also developing were collections of ideas and practices concerned not simply with the shaping of an individual's body and soul but rather with improving and regulating the rhythms of collective populations (Foucault 2003). These bodies of practices, what Foucault termed 'biopolitics', can be discerned most clearly in efforts to understand human pathology and in the concomitant interventions into public hygiene that marked the problematisation of urbanity in the nineteenth century (Gandy 2004; Wood 2005). They can also, as this chapter suggests, be discerned in the biopolitical embrace that was quickly, albeit unevenly, extended to New Zealand's productivist landscapes and the biotic flows that characterised it.

The alienation of Maori land ownership, the clearance of the forests, and importation of flora and fauna to produce a pastoral economy were themselves elements of an enduring biopolitical programme which was overtly designed to 'improve' New Zealand's landscapes. This work was counterpointed by the work of acclimatisation societies (and individuals) which sprang up throughout New Zealand and whose focus was on the introduction and spread of new 'innoxious' species of plants and animals intended to recreate the familiar landscapes of European migrants (Robert 1994). However, the work of individual and acclimatisation societies in introducing new species became increasingly more problematic from the late 19th century onwards. Whilst species such as trout and salmon provided significant recreational (and economic) benefits, other species such as rats, mice, sparrows, wasps, stouts, rabbits, deer and possums quickly became regarded as pests inimical to the good management of a rapidly emerging productivist landscape. Thus, in response to acclimatisation of 'innoxious' species the government was forced to introduce legislation such as the 1882 Small Birds Nuisance Act and the 1884 Codlin Moth Act which empowered local authorities to strike rates to be spent eradicating specifically identified introduced pests. Perhaps the most extensive, and iconically ineffective, efforts at pest control, before the late 20th century focus on possums, were directed at controlling rabbits, beginning with the 1876 Rabbit Nuisance Act in areas such as Otago and Southland in New Zealand's South Island.

Establishing the Department of Agriculture

The ad hoc webs of regulation that developed in response to the threats posed by specific diseases and pests was centralised to some extent with the establishment of the Department of Agriculture in 1892 following the Liberal Party's ascent to power in 1891 (for a detailed account of the Liberal government see Hamer 1988). In founding the new department the Minister of Agriculture, John MacKenzie focused on three objectives: the consolidation of existing agricultural legislation, the centralisation of agricultural administration and increasing the coercive powers of agricultural legislation (Brooking 1996). MacKenzie's centralising objectives can be seen codified in the omnibus 1893 Stock Act (SA 1893) which incorporated provisions concerning sheep, rabbits, diseased cattle and branding. The SA 1893 also centralised inspection powers within the Department of Agriculture and gave stock inspectors very wide ranging powers to enter private property, order the destruction of stock and to fine 'negligent' farmers. The net effect was to significantly strengthen the state's ability to intervene into the work of agricultural producers in order to create a more efficient, competitive industry. On this latter point Mackenzie justified closer state control over activities such as slaughtering as being necessary if New Zealand farmers were to compete with competitors in Australia and the USA (Brooking 1972).

The powers of the Department of Agriculture were further strengthened with the passage of various Dairy Industry Acts in 1892, 1894 and 1898 which established a regime of control that explicitly sought to standardise and improve the quality of New Zealand's dairy exports. The size of the department grew most dramatically, however, with the passage of the 1900 Slaughtering and Inspection Act (SIA 1900). Faced with vehement parliamentary and industry opposition that considered the bill's central principle of comprehensive meat inspection both too expensive and too draconian the bill had a troubled passage into law. But by 1908 the inspection regime that the SIA 1900 had enabled had resulted in the Department of Agriculture's staff increasing from 62 in 1899 to 378 in 1908. An increase primarily made up of new agricultural inspectors.

The growth in inspector numbers was a necessary component in the Department's growing system of policing agricultural producers, but it did not go unnoticed. For example, the future Prime Minister, William Massey joked that there were emerging two classes of colonists, 'those who were Inspectors and those who were inspected' (NZPD 1896: 156); while George Hutchinson railed during the passage of the Dairy Industry Bill that, 'If one-tenth part of the powers proposed to be arrogated to Inspectors by this Bill were attempted to be exercised there would be something like a rising' (NZPD 1898). In relation to this system of inspectors a report produced by the New Zealand Institute of Pacific Relations in the mid-1930s commented that, 'The administration of Statutes containing an element of compulsion necessitates the use of a considerable amount of discretion on the part of responsible officers' (Fawcett 1936: 329), whilst Brooking (1996) suggests that the powers, duties and wide-ranging discretion of the stock inspectors rendered them analogous to a form of rural police.

Policing the Productivist Landscape

This period of regulatory innovation in New Zealand agriculture institutionalised a regime of policing that functioned well into the 20th century. At one scale the inspection system enveloped farmers and their farms in a disciplinary matrix that sought to make the farmers themselves, their animals and indeed landscapes individually visible, and in doing so subject them to a system of norms that would standardise and regularise their practice and production. At wider scale the web of agricultural governance which had its sharp end articulated in the work of the Department of Agriculture's inspectors had a quite overt biopolitical purpose. Though the disciplinary work of inspectors allied with the educational efforts of department instructors and the research work of agricultural scientists the Department of Agriculture sought to know and improve New Zealand's farmers, their farms, their stock, their pests and the network of relationships through which the primary industries created value.

The purpose in undertaking this effort was in the first instance to standardise those relationships, and then to set above improving them. Improvement could take a number of forms. It could mean eliminating sheep diseases such as scabies from the 'national flock', reducing the prevalence of contamination in the meat supply, the genocidal war against rabbits and latterly possums, or eliminating the fishy taint that marked New Zealand's early butter exports. Or it could mean research to improve production: more lambs, more milk, higher yielding crops and so on. The net effect was the production of a system of governance in which the New Zealand state came, to use Torpey's (2000) useful phrase, to 'embrace' farmers and the productivist landscape. Such an embrace profoundly shaped the practice of agriculture in New Zealand. It also, albeit less overtly, significantly transformed the capacity of the New Zealand state to know its various populations, police the rhythms of life, death and productivity in its diverse populations, and to set about improving those populations.

The emergence of the Department of Agriculture, the centralisation of control and the work of its 'rural police' was mirrored in biopolitical programmes being enacted elsewhere in New Zealand. Perhaps the most publically recognisable was the regulatory regime that emerged out the 1890 Sweating Commission which had been charged with investigating allegations that the clothing industry was rife with exploitation and 'sweating'. While the Commission concluded that 'sweating' did not exist in New Zealand's factories it recommended changes to the Factories Act to confer greater powers on industrial inspectors, and the establishment of a system of compulsory industrial arbitration designed to improve New Zealand workplaces (Richardson 1992). Alongside the incorporation of New Zealand's agricultural producers and industrial workers into various regulatory regimes, the country's new towns and cities were increasingly being problematised as sources of disease, pollution and 'old world' sin requiring the provision of clean water, sewerage systems, municipal abattoirs, and controls on pollution and 'slum' dwellings (Wood 2005). In this context the controls that emerged around agricultural production, while specific in their content, were part of a much broader intensification of the New Zealand state's interest in its various populations and landscapes that moved beyond policing *per se* towards a biopolitical concern with improvement.

The biopolitical programme which was institutionalised through the Department of Agriculture evolved to maximise agricultural production, and this focus continued throughout the twentieth century. It was a system of control within which farmers and their farms were closely embraced by agencies such as the Department of Agriculture. However, we should be wary of assuming that there was, and is, a necessary relationship between the ideal of control and the actuality of compliance. Debates over the inspector system, conflict over the allocation of costs for pest eradication, and a questioning of the relevance of the scientific advice coming from the Department of Agriculture allied with a deep suspicion of bureaucratic control from 'Wellington' all suggest that the Department's evolving biopolitical programme was not universally supported (Star and Brooking 2007). On the flip side, however, the sustained growth in agricultural production indicates that while specific elements of the programme may have been contested, its general logic was being supported and adopted.

Neoliberalism and Policing Beyond the State

The biopolitical programme which prioritised production evolved as part of a wider system of agricultural governance which sought to stabilise relations relating to prices, investment, marketing and the environment (Easton 1997). By the 1980s and 1990s these arrangements were under profound pressure and began to be reshaped, along with much wider verities by the application of a remarkably pure version of neoliberalism by the Fourth Labour government (Le Heron and Pawson 1996). Despite these changes there remained ongoing lines of continuity. For example a significant driver of the system that had been assembled during the late nineteenth century concerned the inspection and sanitary certification of New Zealand's meat and dairy exports, and the hygiene concerns which prompted the establishment of this system did not disappear with new modalities of governance. Indeed in their most recent briefing the Ministry of Agriculture and Forestry (2008) has identified increasing consumer concerns about food safety as posing an enduring and fundamental challenge to the productivist ethos of New Zealand's agricultural industries. Nonetheless the roll out of neoliberalism in New Zealand profoundly transformed how farmers and the productivist landscape would be governed.

The neoliberal transformation of New Zealand's state-society relationships was reflected in the 1998 corporatisation of the Ministry of Agriculture and Fisheries's (MAF) commercial activities (MAF Quality Management), and the establishment of ASURE and AgriQuality as state owned enterprises (SOEs) out of the contestable elements of those commercial activities. In announcing the split the then National Government's Acting Minister of Agriculture, John Luxton (NZPD 1998) identified two drivers for it: first, an international move towards a risk management approach to food safety where governments would establish food safety standards and businesses would develop risk management plans to deliver that safety; and second, a belief that MAF Quality Management had used its skills and expertise to expand its activities beyond inspection and that these new areas were not the core business of the government. Speaking more broadly the Acting Minister argued that the corporatisation of the meat inspection system (AgriQuality) and the food safety system (ASURE) would provide increased competition in the inspection and certification market providing an incentive for increased innovation and lower compliance costs. To this end the split would create, 'lower cost structures to ensure that New Zealand's primary producers are internationally competitive long term' (NZPD 1998: 10938).

The new governance structure did not reflect a fundamental deregulation of the biopolitical regime that had emerged around meat inspection and food safety. In other words the enduring concerns with controlling and improving biotic flows that had emerged during the 19th century remained. What was distinctive was the way in which the creation of the two SOEs decentred the state as represented by the Ministry of Agriculture and Food and placed its policing functions into a shifting public-private borderland. In this sense the process is directly analogous to the managerial shift in policing described by Crawford (2003) in which the responsibility for crime prevention has been increasingly dispersed as state agencies redefine the relationship between 'core' and 'non-core' activities. A trajectory which accords with the situation described by Johnston and Shearing (2003) where policing has increasingly given way to the idea of security governance in which the seemingly stable boundary between state and non-state actors dissolves and is replaced in turn with an intricate web of governance nodes, agents and negotiated norms.

The independent existence of ASURE and AgriQuality was revisited in 2007 when the Labour government decided to merge the two SOEs to form a new single SOE -AsureQuality Ltd. A result of what Lewis *et al.* (2008) have described as the gradual questioning by national governments of the neoliberal certainties of the 1980s and 1990s which guided the establishment of ASURE and AgriQuality. In proposing the merger the Minister for State Owned Enterprises, Trevor Mallard (NZPD 2007) reiterated the logic of the initial split but argued that because of the greater demands from consumers for food safety and traceability and the growing risk of biosecurity incursions to the New Zealand economy what was required was an entity with a comprehensive and integrated ability to respond to these challenges. In the creation of a new, single entity, 'Better New Zealand – wider outcomes could result from a more collaborative approach, particularly in the provision of a highly skilled and coordinated standing army, should a major biosecurity threat such as foot-and-mouth or a similar disease strike' (NZPD 2007: 8558). The image of the 'standing army' used by the Minister was striking and belied the seriousness accorded to policing biotic flows into and within New Zealand *vis-à-vis* other economies such as the United Kingdom (at least prior to the 2001 Foot and Mouth outbreak) (see Donaldson and Wood 2004; Jay *et al.* 2003).

The Labour MP Maryan Street (NZPD 2007: 8561) noted the division of labour embedded in the initial split which saw AgriQuality focus on food safety and biosecurity and Asure on the inspection of processed meat, but contended that it had become increasing imperative 'to track live and dead meat in order to be able to determine, in the event of a biosecurity risk, the precise herd from which a piece of processed meat has come, and therefore, to go back to that particular farmer and particular herd and contain any identified risk.' Comments which foreshadowed the introduction in 2008 of a National Animal Identification and Traceability (NAIT) scheme designed to eventually see all stock in New Zealand electronically tagged and traced from 'farm to fork'. Despite criticism from the parliamentary opposition which centred on the cost and anti-competitive nature of the merger, but not the logic of integration, the merger went ahead and a new SOE was formed – AsureQuality Limited – which began trading in October 2007.

AsureQuality Ltd's structure belies the extent to which agricultural producers continued to be enmeshed within a biopolitical embrace that increasingly interweaves concerns about animal and human welfare and stretches from the farm through to markets. Thus they provide on-farm risk management, assurance and inspection services, testing services, biosecurity monitoring and business support all within the general rubric of food safety and biosecurity. (from www.asurequality. com). AsureQuality take pains to highlight the extent of their certification services, including certification services provided as part of the GLOBALGap system. The relationship between AsureQuality Ltd and GLOBALGap neatly reflects the extent to which the biopolitical regime that we began the chapter with has increasingly moved beyond the state's own regulatory apparatus. GLOBALGap emerged in 2007 out of the earlier Euro-Retailer Produce Working Group (EUREP) which had been started by a collection of British retailers and continental European supermarkets in response to growing consumer concerns about food safety, environmental and labour standards (from www.globalgap.org). EUREP represented a private sector initiative to harmonise certification standards as a way of avoiding producers having to undergo multiple audits if they supplied multiple retailers. Within this context GLOBALGap operates as a pre-farm-gate standard which covers farm inputs and all farming activities until the products leave the farm, standards which are enforced through regular inspections via accredited certification bodies such as AsureQuality Ltd in New Zealand.

The significance of GLOBALGap is that it points to the assemblage of global governance regimes which articulate a clear biopolitical interest in policing agricultural production, matched to a disciplinary framework of inspections and audits, but which operate outside state based regulatory systems. The emergence of such systems strongly indicates the continuing embrace of farmers by various forms of biopolitically orientated policing, an embrace, however, that is no longer confined to or indeed necessary shaped by the immediate biopolitical concerns of the state.

Conclusion

This chapter has argued that agriculture in New Zealand has been characterised by the development of systems of policing beyond the police and indeed beyond a singular focus on the human population. This system of surveillance and policing began to emerge during the 19th century in response to a series of biopolitical concerns about the threats to agricultural production caused by a plethora of introduced plants, animals and diseases. The extensive powers of the Department of Agriculture's inspectors represented the first glimmerings of the state's willingness to directly embrace farmers and their farms in a disciplinary matrix as a means of achieving wider biopolitical goals of agricultural and economic improvement.

Whilst the state's desire to embrace agriculture has not diminished, the institutionalisation of that embrace has changed. Here the chapter traced the formation of two particular institutional arrangements – the Department of Agriculture – in the late years of the 19th century and the recasting of that arrangement in the late 20th century with the formation of SOEs in 1998 and 2007. The two institutional arrangements thus described were framed within wider governmental contexts with the formation of the Ministry of Agriculture linked to the emergence of the modern New Zealand state and the formation of the SOEs shaped by the hegemony of a neoliberal orthodoxy that profoundly recast what the New Zealand state would do directly and what it would seek to do indirectly.

In making these arguments this chapter has sought to decentre two assumptions about policing: the singular importance of police as *the* agency of policing; and the position of policing as a core role of the state. In examining rural policing in New Zealand what has been argued is that there are plural policing regimes whose focus extends to embrace wider biotic flows. Whilst policing may well be about the creation and maintenance of social order, in the context of the regimes sketched in this chapter it is a productivist order couched in terms of a biopolitical programme framed by the disordering effects of pests, weeds and disease that has been the focus of sustained attention. Moreover these programmes are no longer confined to the state, rather what has emerged in part as a consequence of the neoliberal transformation of state-society relationships has been the extension of policing not only beyond the police, but also beyond the state. The ongoing implications of this latter shift are both opaque and contingent, but they point to a power geometry which is becoming increasing dispersed, extensive and web like. In this vein this chapter has focused on a series of elite actors, but it needs to be recognised that they are not the only actors involved in an evolving web of control. Much more can be done to explore the biopolitical practices connecting elite actors and more mundane actors. For example, important questions exist over the extent to which individual farmers co-opted and contested the biopolitical regimes being fashioned. Ongoing research should continue exploring the biopolitical webs of policing framing farming because these regimes are constituted and enacted in ways that continue to embrace farmers and productivist landscapes in profound ways.

Chapter 19

W(h)ither Rural Policing? An Afterword

Richard Yarwood and Rob I. Mawby

This book has presented a snapshot of rural policing at the start of the 21st century. Drawing from geography and criminology, it has used a range of theoretical perspectives and empirical examples to examine the policing of rural space across different scales and countries. A number of key themes have emerged from this analysis that have shaped, and will continue to shape, the future of rural policing.

Insiders and Outsiders

A reoccurring theme in many chapters has been the positioning of the police within or outside rural communities and whether policing (by the police and other agencies) should be by, of, or for particular communities (Fyfe and Reeves, Gilling, Mouhanna[1]). The modernisation, rationalisation and neo-liberalisation of police forces have meant the withdrawal of local officers from particular places. The village bobby, who perhaps existed more in myth than reality (Young 1993), has been replaced by reactive policing from a distance and a plethora of community initiatives aimed at 'recreating the best of the village bobby feeling within today's constraints' (West Mercia 2001). As Mouhanna points out, such modernisation has led to losses in contact between the police and rural communities that have been mourned by public and officers alike. Yet, very often this contact benefited selected, elite members of rural communities and contributed to the cultural, rather than criminal, exclusion of others (Barclay *et al.*, Donnermeyer *et al.*, Halfacree, James, Mouhanna, Yarwood). As Woods illustrated, changes in legislation and police relations can lead to a dramatic reversal in who feels excluded from the countryside and its perceived ways of life.

As the term community is increasingly used in rhetoric of rural policing, we would urge critical reflection on what the term means and who is excluded, as well as included, in these measures. It is important to recognise that there are many rural communities, not just one rural community (Gilling). Indeed, rural policing might benefit from the urban experience of policing diverse groups. Such considerations are of increasing importance given changes in the rural policing mix.

---

1 All undated references to authors in this chapter refer to their work in this volume.

Who is Policing?

Many commentators have noted that policing has become 'multilateralised' (Bayley and Shearing 2001: 1) and is no longer the sole responsibility of the state. In neo-liberal countries the last 20 years has seen an expansion in both the number of private agencies involved in policing and the variety of agents employed by different government organisations whose role includes, partially or predominately, a policing responsibility. Work in this book has confirmed that voluntary and, especially, partnership working in a variety of forms (Johnstone; Barton *et al.*; Mawby; Yarwood) are playing an increasingly dominate role in the policing of rural landscapes. Yet, in contrast to many urban spaces (de Waard 1999; Jones and Newburn 1995), private policing has yet, it seems, to make an impact on the rural policing mix (Mawby, Chapter 6). Private security has tended to focus on mass-private spaces, such as shopping malls, or large-scale institutional spaces, including university campuses or industrial sites that are predominantly urban features (Zedner 2009). Although some rural gated communities and institutions, such as military bases, may employ private security guards, the practice appears neither dominant or, on the whole, welcomed by rural public and police. Further systematic research is, however, needed on private security in the policing mix of the countryside to establish its true extent and whether it will, as in urban areas, play a greater role

Not a Happy Lot

This book has aimed to be critical rather criticising. It has presented the experience faced not only by rural people but also the police themselves when policies, practices and the countryside itself changes. In many cases those policing the countryside face a difficult task: one that must balance efficiency against community interaction; local need against national policy; fairness with local sensitivity; and, above all, trying to achieve these over often vast areas with limited resources. Many chapters have also noted that the rural police, in relation to their urban counterparts, have less finances, support and time to achieve their goals. Further, rural policing appears to be withering in many countries as scarce resources are focused on urban places. The police officer's lot is not, therefore, always a happy one.

The authors in this volume have had the luxury of writing and reflecting on policing from their desks rather than the rural front line. While research on rural policing should remain critical, commentators should not lose track of the difficult and sometimes dangerous decisions and actions that officers are required to make. As such, we support Herbert's (1996) call for a theoretical middle ground that considers both the ways in which policing reflects and affects structural inequalities in society (as well as policy changes) yet, at the same time, is sensitive to the ways in which policing is practised in different spaces by those tasked with its implementation. New theorisations of rurality may offer suitable directions here.

Rural Space and Policing

The authors in this volume have taken a range of approaches to conceptualising rurality, ranging from empirical descriptors to theorised interpretations of the term. Although each approach has its own merits, these perspectives offer particular visions that illuminate some aspects of rural policing and hide others. By contrast, Halfacree's (2006a) recent three-fold model of rural space offers an opportunity to triangulate different conceptualisations of rurality to provide a more holistic understanding of the complexities of rural life.

He argues that rural space should be understood in three inter-related ways: first, as a 'rural locality' or physical space that has been shaped by changes in production, consumption and political decision making; second as a 'representation' or image which often reflects elite views but is often challenged by other discourses and voices; and finally through 'everyday lives' or the way that ideas about rurality are performed, embodied and re-produced by different actors. Taken together, these perspectives allow an interrogation of the 'complex interweaving of power relations, social conventions, discursive practices and institutional forces that are constantly combining and recombining' in rural places (Cloke 2006: 24). Thus, to understand policing and rurality it is necessary to consider:

1. *The Rurality Locality.* The sheer scale of rural space has sometimes been forgotten by rural geographers (especially in the UK) yet the chapters on the USA, Australia, Canada and Africa serve to remind us that the distances and isolation faced by many rural police officers are immense and challenging. Equally, it is important to consider the political and structural processes that influence the policing of specific (agricultural or wildlife) crimes (Barclay *et al.*; Fyfe and Reeves) and, significantly, how these have also been shaped themselves by policing (Henry). The physical, social and political environment is more than a backdrop for rural policing; it is both its cause and consequence.

2. *Representations.* To an extent, the countryside and its policing reflect dominate, idyllic visions of rurality (Yarwood and Gardner 2000; Gilling). Groups such as nomads (James; Halfacree), aborigines (Barclay *et al.*) and drug-addicts (Barton *et al.*) are seen to disrupt this ideal and have hence been the target of efforts to police and exclude them from the countryside. Likewise, the victims and survivors of domestic violence remain out of view, this time through a neglect of policing (Squire and Gill). Woods, however, demonstrates that more traditional rural groups, such as farmers and landowners (see also chapters by Fyfe and Reeves, Henry, and Lemieux), have more recently been subject to policing that has enforced particular

visions of the countryside that have criminalised certain (traditional) practices. It is crucial, therefore, to be aware that imaginations of rurality are both complex and contested, not merely out-pourings of the rural idyll. An understanding of the changing social context of the countryside (see above) is thus crucial.

3. *Everyday lives.* Liepins (2000) stresses the importance of everyday actions and performances in the re-production of rural space. These act out representations of rurality, establishing them on the landscape of rural localities. Some chapters have noted that police officers are required not only to interpret the law but also the conventions of rural communities as a well as institutional policing practices in their daily routines (Barclay *et al.*; Fyfe and Reeves; Mouhanna; Yarwood). Such actions may re-enforce particular understandings of community (for example, an officer who does not prosecute a local driver confirms his status as 'belonging' to a community) but also connect rural spaces to national or even international flows of law and governance (enforcing ivory protection, for example, reflects international treaties). More research is, however, needed on the daily practices and performances of policing (Herbert 2006) to establish more clearly the significance of policing (as opposed to simply policy or the law) in shaping rural space.

Beyond Rural Studies

Fyfe (1991: 249) noted that studies of the police have been 'conspicuously absent from the landscapes of human geography'. Studies of rural policing have been even more so, perhaps reflecting that the police themselves are also absent from many rural landscapes. We hope that this volume has gone someway to foregrounding the importance of the police in the countryside to both academic study and the development of policy.

We are, though, realistic enough to note that interest in rural policing remains fickle, in the public eye at least. The conviction of Tony Martin for shooting dead one of two burglars to his isolated Norfolk farm house in 1999 prompted something of a moral panic about rural policing in Britain leading to a plethora of policy initiatives and funding to improve the situation (Yarwood 2008). As the incident has faded from memory and other moral panics, perhaps about the policing of illegal immigration, have emerged, rural policing has moved from centre stage to the wings of policy, practice and research. The situation appears similar in other countries: rural policing gains attention after a major incident only to be relegated in relation to other priorities when panics subside.

As Cloke (2006: 26) notes, one of the tasks of rural research is to ensure that ideas are 'made' and not simply deployed from other areas of research. Thus, rigorous, sustained and critical research on rural policing will ensure that it retains a presence in not only the landscapes of geographical and criminological study but also in those of policy makers and the police. Only then will rural research create tracks of its own, rather than following belatedly in those made by other researchers or policy makers.

Bibliography

Aboriginal Policing Directorate 2007. *Aboriginal Policing Update*. [Online: Public Safety Canada]. Available at: http://www.publicsafety.gc.ca/prg/le/ap/_ fl/update_vol01_no01_e.pdf [Accessed 9th May 2010].

Aboriginal and Torres Strait Islander Women's Task Force on Violence 2000. *Final Report*. Brisbane: DATSIPD.

ACPO 2005. *The Hunting Act 2004: National Tactical Considerations. Note to ACC Operations*. London: Association of Chief Police Officers.

ACPO 2008. *Guidance on Unauthorised Encampments*. London: Association of Chief Police Officers.

ACPOS 2006. *A Strategy for Dealing with Wildlife Crime in Scotland*. Glasgow: Association of Chief Police Officers, Scotland.

Adcock, G. 2002. *Country Policing: It's not Rocket Science*. Canberra: Winston Churchill Memorial Trust of Australia.

Alsop, R., Clisby, S., Craig, G., Evans, R. and Hockey, J. 2002. *Beyond the Bus Shelter: Young Women's Choices and Challenges in Rural Areas*. Oxford: YWCA.

Ames, W. 1981. *Police and Community in Japan*. Berkeley: University of California Press.

Anderson, J. and Tresidder, J. 2008. *A Review of the Western Australian Community Safety and Crime Prevention Partnership Planning Process*. Perth: Australian Institute of Criminology.

Anderson, K. and McCall, M. 2005. *Farm Crime in Australia*. Canberra: Australian Institute of Criminology.

Anderson, S. 1997. *Crime in Rural Scotland*. Edinburgh: The Scottish Office.

Anderson, S. 1999. Crime and Social Change in Rural Scotland, in *Crime and Conflict in the Countryside*, edited by Dingwall, G. and Moody, S. Cardiff: University of Wales Press, 45–59.

Anderton, K. 1985. *The Effectiveness of Home Watch Schemes in Cheshire*. Chester: Cheshire Police.

Animal Health Australia 2008. *National Livestock Identification System*. [Online]. Available at: http://www.animalhealthaustralia.com.au/programs/adsp/nlis/ nlis_home.cfm [Accessed 3rd September 2008].

Armstrong, N. 2005. *The Impact of Crime in Rural Fermanagh*. Belfast: Queen's University, M.S. Thesis, April.

Aust, R. and Simmons, J. 2002. *Rural Crime: England and Wales. Home Office Statistical Bulletin 01/02*. London: Home Office.

Australian Bureau of Statistics 2006. *Population Distribution, Aboriginal and Torres Strait Islander Australians, 2006, 4705.0.* Canberra: Australian Bureau of Statistics.

Australian Bureau of Statistics 2008. *Australian Social Trends, 2008, 4102.0.* Canberra: Australian Bureau of Statistics.

Baker, D. 2007. Policing industrial conflict in rural and regional settings: local and 'outside' approaches. *International Journal of Rural Crime*, 1, 79–92.

Balding, J. 1998. *Young People and Illegal Drugs in 1998.* Exeter: Schools Health Education Unit.

Bancroft, A. 2000. "No interest in land": legal and spatial enclosure of gypsy-travellers in Britain. *Space and Polity*, 4, 41–56.

Bancroft, A. 2005. *Roma and Gypsy-Travellers in Europe: Modernity, Race, Space and Exclusion.* Avebury: Ashgate.

Bannister, J., Fyfe, N. and Kearns, A. 1998. Closed circuit television and the city, in *Surveillance, Closed Circuit Television and Social Control*, edited by Norris, C. *et al.* Aldershot: Ashgate, 21–39.

Barclay, E. and Donnermeyer, J.F. 2002. Property crime and crime prevention on farms in Australia, *Crime Prevention and Community Safety: An International Journal*, 4, 47–61.

Barclay, E. and Donnermeyer, J.F. 2007. Farm victimisation: the quintessential rural crime, in *Crime in Rural Australia*, edited by Barclay, E.M. *et al.* Sydney: Federation Press, 57–68.

Barclay, E. and Donnermeyer, J.F. 2009. Crime and security on agricultural operations. *Security Journal*, 21, 1–18.

Barclay, E., Donnermeyer, J., Doyle, B. and Tarlay, D. 2001. *Property Crime Victimisation and Crime Prevention on Farms.* Armidale: Institute for Rural Futures, University of New England.

Barclay, E., Scott J., Donnermeyer, J., Hogg, R (eds) 2007. *Crime in Rural Australia.* Sydney, Australia: Federation Press.

Barnard, M. and Forsyth, A. 1998. Drug use among school children in rural Scotland. *Addiction Research*, 6:5, 421–34.

Barnes, R., Craig, G., Dublin, H., Overton, G., Simons, W. and Thouless, C. 1999. *African Elephant Database 1998. Report Number 22.* Glad, Switzerland: African Elephant Specialist Group, IUCN Species Survival Commission.

Barr, R. and Pease, K. 1992. The problem of displacement, in *Crime, Policing and Place: Essays in Environmental Criminology*, edited by Evans, D. *et al.* London: Routledge and Kegan Paul.

Barton, A. 2005. Working in the margins: shadowland agencies, outreach workers and the crime audit process. *Drugs: Education, Prevention and Policy*, 12:3, 239–46.

Barton, A. and James, Z. 2003. 'Run to the sun': policing contested perceptions of risk. *Policing and Society*, 13, 259–70.

Bayley, D.H. 1976. *Forces of Order: Police Behaviour in Japan and the United States.* Berkeley: University of California Press.

Bayley, D.H. 1985. *Patterns of Policing.* New Brunswick: Princetown University Press.

Bayley, D.H. 1994. *Police for the Future.* Oxford: Oxford University Press.

Bayley, D.H. and Shearing, C. 2001. *The New Structure of Policing: Description, Conceptualization and Research Agenda.* Washington, DC: US Department of Justice National Institute of Justice Research Report.

Bayley, R.I. 1986. *Community Policing in Australia: An Appraisal.* Payneham: National Police Research Unit.

BBC News 2008. 'Big Rise' in England Farm Thefts, 4th September, 2008. [Online: BBC] Available at: http://news.bbc.co.uk/1/hi/uk/7553538.stm [Accessed 15th February 2009].

BBC News 2008. Sea Eagles Blamed for Lamb Deaths. [Online: BBC] Available at: http://news.bbc.co.uk/1/hi/uk/7630655.stm [Accessed 23rd September 2008].

Bean, T.L. and Lawrence, L.D. 1978. *Crime on Farms in Hampshire County, West Virginia.* Morgantown: College of Agriculture and Forestry, West Virginia University.

Beck, A. and Willis, A. 1995. *Crime and Security: Managing the Risk to Safe Shopping.* Leicester: Perpetuity Press.

Beier, A. 1985. *Masterless Men. The Vagrancy Problem in England 1560–1640.* London: Methuen.

Beirne, P. and South, N. (eds) 2007. *Issues in Green Criminology.* Cullompton: Willan Publishing.

Belich, J. 1996. *Making Peoples: A History of the New Zealanders from Polynesian Settlement to the End of the Nineteenth Century.* Auckland: Allen Lane.

Bell, D. 2006. Variations on the rural idyll, in *Handbook of Rural Studies*, edited by Cloke, P. *et al.* Sage: London, 149–60.

Bennett, T. 1990. *Evaluating Neighbourhood Watch.* Aldershot: Gower.

Bergère, M. 2004. Épouser un gendarme ou épouser la gendarmerie? Les femmes de gendarmes entre contrôle matrimonial et contrôle social, *Clio, revue d'histoire*, 20, 123–34.

Blagg, H. 2003. *An Overview of Night Patrol Services in Australia.* Canberra: Attorney-General's Department.

Blagg, H. and Valuri, G. 2004. Aboriginal community patrols in Australia: self–policing, self–determination and security. *Policing and Society*, 14:4, 313–28.

Blanc, J., Barnes, R.F.W., Craig, G.C., Dublin, H.T., Thouless, C.R., Douglas-Hamilton, I. and Hart, J.A. 2007. *African Elephant Status Report 2007: An Update from the African Elephant Database. Report Number 33.* Glad, Switzerland: African Elephant Specialist Group, IUCN Species Survival Commission.

Blanc, J., Thouless, C., Hart, J., Dublin, H., Douglas-Hamilton, I., Craig, C. and Barnes, R. 2003. *African Elephant Status Report 2002. Report Number 29.* Glad, Switzerland: African Elephant Specialist Group, IUCN Species Survival Commission.

Blight, T. 2004. *OCP Presentation to Shire Councils.* Perth: Office of Crime Prevention.

Bone, V. 2008. 'We have to fortify our fields' [On-line: *BBC News* 4th September] Available at: http://news.bbc.co.uk/1/hi/uk/7484909.stm [accessed: April 2009].

Bouza, A.V. 1990. *The Police Mystique: An Insider's Look at Cops, Crime and the Criminal Justice System.* New York: Plenum.

Bowling, B. and Foster, J. 2002. Policing and the police, in *The Oxford Handbook of Criminology Third Edition*, edited by Maguire, M. *et al.* Oxford: Oxford University Press, 980–1033.

Bowling, B. and Phillips, C. 2002. *Racism, Crime and Justice.* Harlow: Longman.

Boyle, T. 2007. Small towns have higher crime rates. *Toronto Star*, 29th June, A1.

Brå 2008. *Poaching for Large Predators – Conflict in a Lawless Land?* Stockholm: Brå Swedish National Crime Prevention Council.

Bradford, B., Jackson, J. and Stanko, E.A. 2009. Contact and confidence: revisiting the impact of public encounters with the police. *Policing and Society*, 19, 1: 20–46.

Bradshaw, R., Cuff, J., Rogers, J. and Watkins, L. 2006. *Rural Disadvantage: Reviewing the Evidence.* [Online: Commission for Rural Communities]. Available at: http://www.ruralcommunities.gov.uk/files/CRC31-RuralDisadvantage-reviewingtheevidence.pdf [Accessed 7th October 2008].

British Columbia's Police Act 1996. *Police Act, Revised Statutes of British Columbia 1996.* Victoria, British Columbia: Queen's Printer, Ministry of Finance and Corporate Relations.

Brooking, T. 1972. *Sir John Mackenzie and the Origins and Growth of the Department of Agriculture 1891–1900.* Palmerston North: Unpublished MA Thesis, Department of History, Massey University.

Brooking, T. 1996. *Lands for the People? The Highland Clearances and the Colonisation of New Zealand: A Biography of John McKenzie.* Dunedin: University of Otago Press.

Brooking, T. and Pawson, E. 2007. Silences of grass: retrieving the role of pasture plants in the development of New Zealand and the British Empire. *Journal of Imperial and Commonwealth History*, 35:3, 417–35.

Brown, D. 2009. Defiant travellers facing eviction say they'll move – to the next field. *The Times.* 23rd January.

Brown, J. and Young, C. 1995. *Substance Misuse in Rural Areas.* London: NCVO.

Brown, P. and Niner, P. 2009. *Assessing Local Housing Authorities' Progress in Meeting the Accommodation Needs of Gypsy and Traveller Communities in England.* London: EHRC.

Brown, P.L. 2004. Someone is stealing avocados, and 'Guac Cops' are on the case. The *New York Times.* 26th January, A1.

Brown, S. 1998. What's the problem, girls? CCTV and the gendering of public safety, in *Surveillance, Closed Circuit Television and Social Control*, edited by Norris, C. *et al.* Aldershot: Ashgate, 207–20.

Bryant, B. 1996. *Twyford Down: Roads, Campaigning and Environmental Law.* London: Spon.

Bryne, R. 2007. Horizon scanning rural crime – agroterrorism an emerging threat to UK agriculture? *International Journal of Rural Crime*, 1:1, 62–78.

Bucke, T. and James, Z. 1998. *Trespass and Protest: Policing Under the Criminal Justice and Public Order Act 1994.* London: Home Office.

Bullock, S. 2008. *Police Strength Statistics, England and Wales*, 31 March 2008 Second Edition. London: Home Office Statistical Bulletin.

Bureau of Justice Statistics 2002. *Sourcebook of Criminal Justice Statistics.* [Online: Bereau of Justice Statistics]. Available at: www.albany.edu/sourcebook [Accessed 12 December 2008].

Bureau of Justice Statistics 2008. *Sourcebook of Criminal Justice Statistics.* [Online: Bereau of Justice Statistics]. Available at: www.albany.edu/sourcebook [Accessed 12 December 2008].

Butterworth, S. 2005. *More than Law and Order: Policing a Changing Society, 1945–1992.* Dunedin: University of Otago Press.

Buttle, J. 2006. *What is Known About Policing Rural Crime: Reviewing the Contemporary Literature. Report to New Zealand Police Rural Liaison.* Auckland: Auckland University of Technology.

Cain, M. 1973. *Society and the Policeman's Role.* London: Routledge and Kegan Paul.

Callinan, I. 2008. *Equine Influenza: The August 2007 Outbreak in Australia, Report to the Equine Influenza Inquiry.* [Online: Australian Government]. Available at: http://www.equineinfluenzainquiry.gov.au/ [Accessed 20th February 2009].

Campbell, S. 1995. Gypsies: the criminalisation of a way of life? *Criminal Law Review.* January, 28–37.

Carcach, C. 2002. *Farm Victimisation in Australia. No. 235 in Trends and Issues in Crime and Criminal Justice.* Canberra, AU: Australian Institute of Criminology.

Carraz, R. and Hyest, J. 1998. *Une Meilleure Répartition des Effectifs de la Police et de la Gendarmerie pour une Meilleure Sécurité Publique: Rapport Au Premier Ministre.* Paris: La documentation Française.

Carrington, K. 2007. Violence and the architecture of rural life, in *Crime in Rural Australia*, edited by Barclay, E. *et al.* Sydney: Federation Press, 88–99.

Carrington, K. and Scott, J. 2008. Masculinity, rurality and violence. *British Journal of Criminology*, 48. 3, 641–66.

Carter, B. and MacGibbon, J. 2003. *Wool: A History of New Zealand's Wool Industry.* Wellington: Ngaio Press.

Caulkins, J. and Reuter, P. 1998. What the price data tells us about drug markets. *Journal of Drugs Issues*, 28: 3, 593–612.

Cemlyn, S., Greenfields, M., Burnett, S., Matthews, Z. and Whitwell, C. 2009. *Inequalities Experienced by Gypsy and Traveller Communities: A Review.* London: EHRC.

Chaiken, M.R. 2004. *Community Policing Beyond the Big Cities.* [Online: COPS, US Department of Justice, National Institute of Justice]. Available at: www.ncjrs.gov/pdffiles1/nij/205946.pdf [Accessed 2nd July 2009].

Chakraborti, N. and Garland, J. 2003. An 'invisible' problem? Uncovering the nature of racial victimisation in rural Suffolk. *International Review of Victimology*, 10, 1–17.

Chakraborti, N., and Garland, J. (eds) 2005. *Rural Racism.* Cullompton: Willan.

Chalfin, A., Roman, J., Mears, D. and Scott, M. 2007. *The Costs and Benefits of Agricultural Crime Prevention.* Washington DC: The Urban Institute.

Chan, J. 2003. Police and new technologies, in *Handbook of Policing*, edited by Newburn, T. Cullompton: Willan, 655–79.

Chapman, B. 1978. The Canadian police: a survey. *Police Studies*, 1.1, 62–72.

Chivite-Matthews, N., Richardson, A., O'Shea, J., Becker, J., Owen, N., Roe, S. and Condon, J. 2005. *Drug Misuse Declared: Findings from the 2003/04 British Crime Survey England and Wales.* London: Home Office Statistical Bulletin.

CITES 2004. *CoP13 Doc. 29.3. Annex 9 Preliminary Information on Elephant Poaching in Regard to the MIKE Central Africa Forest Surveys.* [Online: CITIES]. Available at: http://www.cites.org/common/cop/13/inf/E13-29-3-A9.pdf [Accessed 14th April 2010].

CITES 2008. *Convention on International Trade in Endangered Species of Wild Flora and Fauna.* [Online: CITIES]. Available at: http://www.cites.org [Accessed 14th April 2010].

Clark, A. 1949. *The Invasion of New Zealand by People, Plants and Animals: The South Island.* New Brunswick: Rutgers University Press.

Clark, C. and Greenfields, M. 2006. *Here to Stay: The Gypsies and Travellers of Britain.* Hatfield: University of Hertfordshire Press.

Clayton, M. 2004. *Endangered Species: Foxhunting – the History, the Passion and the Fight for Survival.* Shrewsbury: Swan Hill.

Cleach, O. 2007. *La Désobéissance dans une Organisation D'ordre, L'exemple Du Conflit des Gendarmes, de Décembre 2001.* Unpublished Thesis Paris: Université de Paris – IX Dauphine.

Cleland, C.L. 1990. *Crime and Vandalism on Farms in Tennessee: Farmer Opinions About and Experiences With.* Knoxville: Agricultural Experiment Station, Institute of Agriculture, University of Tennessee.

Clément, S. 2003. *Vivre en Caserne à L'Aube du XXIe Siècle. L'Exemple de la Gendarmerie.* Paris: L'Harmattan, Collection Le travail du social.

Cloke, P. 2006. Conceptualising rurality, in *Handbook of Rural Studies*, edited by Cloke, P. London: Sage, 18–27.

Cloke, P. and Goodwin, M. 1992. Conceptualising countryside change: from post-fordism to rural structured coherence. *Transactions of the Institute of British Geographers, New Series*, 17, 321–36.

Cloke, P., Goodwin, M., Milbourne, P. and Thomas, C. 1995. Deprivation, poverty and marginalisation in rural lifestyles in England and Wales. *Journal of Rural Studies*, 11:4, 351–65.

Cloke, P., Goodwin, M. and Mooney, P. (eds) 2006. *Handbook of Rural Studies.* London: Sage.

Cloke, P. and Little, J. (eds) 1997. *Contested Countryside Cultures. Otherness, Marginalisation and Rurality.* London: Routledge.

Cloke, P., Milbourne, P. and Widdowfield, R. 2001. Homelessness and rurality: exploring connections in local spaces of rural England. *Sociologia Ruralis*, 41:4, 438–53.

Clout, H. 1972. *Rural Geography.* Oxford: Pergamon.

Cocklin, C., Walker, L. and Blunden, G. 1999. Cannabis highs and lows: sustaining and dislocating rural communities in Northland, New Zealand. *Journal of Rural Studies*, 15:3, 241–55.

Cohen, S. 1972. *Folk Devils and Moral Panics.* London: MacGibb and Kee.

Commission for Racial Equality 2006. *Common Ground: Equality, Good Race Relations and Sites for Gypsies and Irish Travellers.* London: Commission for Racial Equality.

Commission for Rural Communities 2006a. *Reviewing the Work of Parish and Town Councils in Reducing Crime.* Cheltenham: Commission for Rural Communities.

Commission for Rural Communities 2006b. *Parish and Town Councils Working Towards Crime Prevention: An Evaluation and Good Practice Paper.* Cheltenham: Commission for Rural Communities.

Condon, J. and Smith, N. 2003. *Prevalence of Drug Use: Key Findings from the 2002/2003 British Crime Survey. Findings 229.* London: Home Office.

Conway, E. 1999. *The Recording of Wildlife Crime in Scotland.* Edinburgh: Scottish Office.

Cooper, C., Anscombe, J., Avenell, J., McLean, F. and Morris, J. 2006. *A National Evaluation of Community Support Officers.* London: Home Office Home Office.

Coorey, L. 1990. Policing of violence in rural areas, in *Rural Health and Welfare in Australia*, edited by Cullen, T. *et al.* Centre for Rural Welfare Research: Charles Sturt University, Wagga Wagga, 80–101.

Coquilhat, J. 2008. *Community Policing: An International Literature Review.* Wellington: New Zealand Police.

Corboy, M., Gilbert, E., Purdie, R. and McKenna, K. 2005. *Communication Centres Service Centre Independent External Review.* Wellington: Police National Headquarters, New Zealand Police.

Cornish, D. and Clarke, R. 2003. Opportunities, precipitators and criminal decisions: a reply to wortley's critique of situational crime prevention, in *Theory*

for Practice in Situational Crime Prevention, Crime Prevention Studies, edited
by Smith, M. and Cornish, D. Monsey, NY: Criminal Justice Press, 41–96.

Countryside Agency 2004. *The State of the Countryside 2004*. London: The
Countryside Agency.

Countryside Agency and Save the Children 2003. *Children and Domestic Violence
in Rural Areas*. Plymbridge: Save the Children.

Countryside Online 2004. *Rural Crime Perceived to be on the Increase*. [Online:
National Farmer's Union]. Available at: www.countrysideonline.co.uk/
newsfarming-1055.htm [Accessed 12th July 2004].

Coupe, T. and Griffiths, M. 1997. *Solving Residential Burglary. Crime Detection
and Prevention Series, Paper no.77*. London: Home Office.

Courouble, M., Hurst, F. and Milliken, T. 2003. *More Ivory than Elephants:
Domestic Ivory Markets in Three West African Countries*. Cambridge:
TRAFFIC.

Cowlishaw, G. 1987. Policing of races in rural Australia. *Journal of Studies in
Justice*, 11, 43–52.

Coy, M., Lovett, J. and Foord, J. 2009. *Maps of Gaps: The Post-code Lottery of
Violence Against Women Support Services*. London: End of Violence Against
Women.

Cragg Ross Dawson 2003. *Drugs in Rural Areas. Qualitative Research to Assess
the Adequacy of Communications and Services*. London: Home Office.

Crawford, A. 1997. *The Local Governance of Crime: Appeals to Community and
Partnerships*. London: Clarendon Press.

Crawford, A. 1998. *Crime Prevention and Community Safety*. Harlow: Longman.

Crawford, A. 2003. The pattern of policing in the UK: policing beyond the police,
in *Handbook of Policing*, edited by Newburn, T. Cullompton: Willan, 136–
169.

Crawford, A., Lister, S., Blackburn, S. and Burnett, J. 2005. *Plural Policing:
the Mixed Economy of Visible Patrols in England and Wales*. Bristol: Policy
Press.

Cresswell, T. 1996. *In Place/Out of Place*. Minneapolis: University of Minnesota
Press.

Cross, P. 2007. Femicide: Violent partners create war zone for women. *Toronto Star*,
6th July, AA8.

Crowther, M. 1992. The tramp in *Myths of the English*, edited by Porter, R.
Cambridge: Polity, 91–113.

Cullingworth, B. and Nadin, V. 2006. *Town and Country Planning in the UK 14th
Edition*. London: Routledge.

Cunneen, C. 1992. Policing and Aboriginal communities: is the concept of over-
policing useful? in *Aboriginal Perspectives on Criminal Justice*, edited by
Cunneen, C. Sydney: Sydney University Institute of Criminology, 76–92.

Cunneen, C. 2001. *Conflict Politics and Crime: Aboriginal Communities and the
Police*. Sydney: Allen and Unwin.

Cunneen, C. 2005. *Evaluation of the Queensland Aboriginal and Torres Strait Islander Justice Agreement. Report to the Justice Agencies CEOs, Brisbane.* Unpublished.

Cunneen, C. 2006. Racism, discrimination and the over-representation of indigenous people in the criminal justice system: some conceptual and explanatory issues. *Current Issues in Criminal Justice*, 17:3, 329–46.

Cunneen, C. 2007. Crime, justice and indigenous people, in *Crime in Rural Australia*, edited by Barclay, E. *et al.* Leichhardt: The Federation Press, 142–53.

Daily Telegraph 1997. A Spy in the Village Square. 25th January.

Davidson, N., Sturgen-Adams, L. and Burrows, C. 1997. *Tackling Rural Drug Problems: A Participatory Approach. Crime Detection and Prevention Series Paper 81, Police Research Group.* London: Home Office.

Davis, J. 1997. New Age Travellers in the countryside: incomers with attitude, in *Revealing Rural 'Others'*, edited by Milbourne, P. London: Pinter, 117–34.

Dawson, R. 2000. *Crime and Prejudice: Traditional Travellers.* Derbyshire: Blackwell.

de Swaan, A. 1995. Widening circles of identification: emotional concerns in sociogenetic perspective. *Theory, Culture and Society*, 12, 25–39.

de Waard, J. 1999. The private security industry in international perspective. *European Journal on Criminal Policy and Research*, 7, 143–74.

Deane, M. and Doran, S. 2002. *Section 17 of the Crime and Disorder Act 1998: A Practical Guide for Parish and Town Councils.* London: NACRO/Countryside Agency.

Decker, S. 1979. Rural county sheriff: an issue in social control. *Criminal Justice Review*, 4.2, 97–111.

Deeds, J., Frese, W., Hitchner, M. and Solomon, M. 1992. *Farm Crime in Mississippi.* Mississippi State: Mississippi State University.

DeKeseredy, W.S. 2009. Canadian crime control in the new millennium: The influence of new-conservative U.S. policies and practices. *Police Practice and Research*, Vol. 10, 305–11.

DeKeseredy, W.S., Alvi, S., Schwartz, M. 2006. Left realism revisited, in *Advancing Critical Criminology*, edited by DeKeseredy, W. and Perry, B. Lanham, Massachusetts: Lexington Books, 19–42.

DeKeseredy, W.S., Alvi, S., Schwartz, M.D. and Tomaszewski, E.A. 2003. *Under Siege: Poverty and Crime in a Public Housing Community.* Lanham, MD: Lexington Books.

DeKeseredy, W.S., Donnermeyer J., Schwartz, M., Tunnell, K. and Hall, M. 2007. Thinking critically about rural gender relations: toward a rural masculinity crisis/male peer support model of separation/divorce sexual assault. *Critical Criminology*, 15, 295–311.

DeKeseredy, W.S., Donnermeyer J. and Schwartz, M., 2009. Toward a gendered second generation CPTED for preventing woman abuse in rural communities. *Security Journal.*

DeKeseredy, W.S. and Joseph, C. 2006. Separation/divorce sexual assault in rural Ohio: Preliminary results of an exploratory study. *Violence Against Women*, 12, 301–11.

DeKeseredy, W.S. and MacLean, B. 1993. Critical criminological pedagogy in canada: strengths, limitations, and recommendations for improvements. *Journal of Criminal Justice Education*, 4, 361–76.

Deleuze, G. and Guattari, F. 1987. *A Thousand Plateaus*. Minneapolis: University of Minnesota Press.

Department of Environment, Transport and the Regions 2000. *Our Countryside: The Future: A Fair Deal for Rural England*. London: HMSO.

Desplanques, G. 2005. Géographie de la population âgée en France : Les nouvelles données démographiques. *Retraite et Société*, 45, 23–41.

Dingwall, G. and Moody, S. (eds) 1999. *Crime and Conflict in the Countryside*. Cardiff: University of Wales Press.

Dion, R. 1982 *Crimes of the Secret Police*. Montreal: Black Rose Books.

Director-General of Agriculture 1945. Department of Agriculture: Annual Report for 1944–45. *Appendix to the Journal of the House of Representatives*, V. III: H-29.

Ditton, J., Short, E., Phillips, S., Norris, C. and Armstrong, G. 1999. *The Effect of Closed Circuit Television Cameras on Recorded Crime Rates and Public Concern about Crime in Glasgow*. Edinburgh: Scottish Office Central Research Unit.

Doherty, B., Paterson, M., Plows, A. and Wall, D. 2003. Explaining the fuel protests. *British Journal of Politics and International Relations*, 5, 165–73.

Donaldson, A. and Wood, D. 2004. Surveilling strange materialities: categorisation in the evolving geographies of FMD biosecurity. *Environment and Planning D: Society and Space*, 22:3, 373–91.

Donnermeyer, J.F. 1987. *Crime Against Farm Operations*. Columbus: Department of Agricultural Economics and Rural Sociology, Ohio State University.

Donnermeyer, J.F. 1994. Crime and violence in rural communities, in *Perspectives on Violence and Substance Use in Rural America*, edited by Blaser, J. Chicago, Illinois: Midwest Regional Center for Drug-Free Schools and Communities and the North Central Regional Educational Laboratory.

Donnermeyer, J.F. 2007a. Locating rural crime: the role of theory, in *Crime in Rural Australia*, edited by Barclay, E. *et al*. Sydney. Australia: Federation Press, 15–26.

Donnermeyer, J.F. 2007b. Rural crime: roots and restoration. *International Journal of Rural Crime*, 1:1, 1–20.

Donnermeyer, J.F. and Barclay, E. 2005. The policing of farm crime. *Police Practices and Research*, 6:1, 3–17.

Donnermeyer, J.F. and DeKeseredy, W. 2008. Toward a rural critical criminology. *Southern Rural Sociology*, 23:2, 4–28.

Donnermeyer, J., Jobes, P. and Barclay, E. 2006. Rural crime, poverty, and community, in *Advancing Critical Criminology*, edited by DeKeseredy, W. and Perry, B. Lanham: Lexington Books, 199–218.

Donnermeyer, J.F. and Tunnell, K. 2007. *In Our Own Backyard: Methamphetamine Production, Trafficking and Abuse in Rural America. Rural Reality Series of Policy Briefs. Number 6*. Columbia, MO.: Rural Sociological Society.

Douglas, M. 1992. *Purity and Danger: An Analysis of Concepts of Pollution and Taboo*. Routledge, London.

Drummond, I., Campbell, H., Lawrence, G. and Symes, D. 2000. Contingent or structural crisis in British agriculture?. *Sociologia Ruralis*, 40, 111–27.

Dunkelberger, J.E., Clayton, J.M., Myrick, R.S. and Lyles, G.J. 1992. *Crime and Alabama Farms: Victimization, Subjective Assessment, and Protective Action*. Auburn, AL: Auburn University.

Dunstall, G. 1999. *A Policeman's Paradise? Policing a Stable Society, 1918–1945*. Palmerston North: The Dunmore Press.

Durkheim, E. 1893. *De la Division du Travail Social*. Paris: PUF.

Eapinoza, E. and Mona, M. 1999. *Identification Guide for Ivory and Ivory Substitutes*. [Online: CITES]. Available at: http://www.cites.org/eng/resources/pub/E-Ivory-guide.pdf [Accessed January 2009].

Easterbrook, S., Yu, E., Aranda, J., Fan, Y., Horkoff, J., Leica, M. and Quadir, R. 2005. *Do Viewpoints Lead to Better Conceptual Models? An Exploratory Case Study: Proceedings of the 13th IEEE International Conference on Requirements Engineering RE'05*. Washington: IEEE Computer Society, 199–208.

Eastern Kentucky University 2003. *National Assessment of Technology and Training for Small and Rural Law Enforcement Agencies NATTS: A Descriptive Analysis*. [Online: US Department of Justice]. Available at: www.ncjrs.gov/pdffiles1/nij/grants/198619.pdf [Accessed 2nd July 2009].

Easton, B. 1997. *In Stormy Seas: The Post-War New Zealand Economy*. Dunedin: University of Otago Press.

Edwards, M. 2008. *NT Intervention Head Says More Needs to be Done*. [Online: Australian Broadcasting Commission]. Available at: http://www.abc.net.au/cgibin/common/printfriendly.pl?http://www.abc.net.au/am/content/2008/s2270854.htm [Accessed June 2008].

Edwards, S. 1989. *Policing Domestic Violence*. London: Sage.

Emsley, C. 1983. *Policing and its Context, 1750–1870*. London: Macmillan.

Emsley, C. 1999. *Gendarmes and the State in Nineteenth-Century Europe*. Oxford: Oxford University Press.

Environmental Audit Committee 2004. *Environmental Crime: Wildlife Crime: Twelfth Report of Session 2003–04*. London: House of Commons/The Stationery Office.

Ericson, R. 1982. *Reproducing Order: A Study of Police Patrol Work*. Toronto: University of Toronto Press.

Erikson, K. 1962. Notes on the sociology of deviance. *Social Problems*, 9, 307–314.

Evans, B. 1969. *A History of Agricultural Production and Marketing in New Zealand.* Palmerston North: Keeling and Mundy.

EVAW 2008. *Making the Grade? The Third Annual Independent Analysis of Government Initiatives on Violence Against Women.* London: Amnesty International UK.

Fahchamps, M. and Moser, C. 2003. Crime, isolation and law enforcement. *Journal of African Economics*, 12, 625–71.

Faivre, J. 1993. Le travail des policiers en tenue dans un commissariat central parisien, in *Police, Justice, Prison, Trois Études de Cas*, edited by Ackermann, W. Paris: L'Harmattan, 65–123.

Falcone, D.N. and Wells, E.L. 1995. The county sheriff as a distinctive policing modality, *American Journal of Police*, XIV:3/4, 123–49.

Falcone, D.N., Wells, E. and Weisheit, R. 2002. The small-town police department, *Policing: An International Journal of Police Strategies and Management*, 25:2, 371–84.

Farmer, F.L. and Voth, D.E. 1989. *Ecological Characteristics of Farm Victimization in Arkansas.* Fayetteville: Arkansas Agricultural Experiment Station, University of Arkansas.

Fawcett, E. 1936. The Department of Agriculture, in *Agricultural Organisation in New Zealand: A Survey of Land Utilization, Farm Organisation Finance and Marketing*, edited by Williams, D. Melbourne: Melbourne University Press.

Federated Farmers of New Zealand. 2008. *2008 General Election Manifesto: New Zealand's Economic Backbone.* Wellington: Federated Farmers of New Zealand.

Finnane, M. (ed.) 1987. *Policing in Australia: Historical Perspectives.* Sydney: NSW University Press.

Fiske, J. 1998. Surveilling the city. *Theory, Culture and Society*, 152, 67–88.

Fitzgerald, J. and Weatherburn, D. 2001. *Aboriginal Victimisation and Offending: The View from Police Records, Crime and Justice Statistics Brief.* Sydney: New South Wales Bureau of Crime Statistics and Research.

Flood-Page, C., Campbell, S., Harrington, V. and Miller, J. 2000 *Youth Crime: Findings from the 1998/99 Youth Lifestyle Survey. Home Office Research Study 209.* London: Home Office.

Foucault, M. 1977. *Discipline and Punish: The Birth of the Prison.* London: Penguin Books.

Foucault, M. 1991. Governmentality, in *The Foucault Effect: Studies in Governmental Rationality*, edited by Burchell, G. and Miller, P. Hemel Hempstead: Harvester Wheatsheaf, 87–104.

Foucault, M. 2003. *Society Must Be Defended: Lectures at the College de France, 1975–1976.* New York: Picador.

Freudenburg, W.R. 1986. The density of acquaintanceship: an overlooked variable in community research?, *American Journal of Sociology*, 92, 27–63.

Friends, Families and Travellers 2009. *Homepage*. [Online: Friends, Families and Travellers]. Available at: http://www.gypsy-traveller.org/ [Accessed April 2009].

Fyfe, N. 1991. The police, space and society: the geography of policing. *Progress in Human Geography*, 15, 249–67.

Fyfe, N. 1995. Law and order policy and the spaces of citizenship in contemporary Britain. *Political Geography*, 14, 177–89.

Fyfe, N. and Bannister, J. 1996. City watching: closed circuit television in public spaces. *Area*, 281, 37–46.

Gagne, P.L. 1992. Appalachian women: violence and social control. *Journal of Contemporary Ethnography*, 20, 387–415.

Galliher, J., Donavan, L.P. and Adams, D.L. 1975. Small-town police: trouble, tasks, and publics. *Journal of Police Science and Administration*, 3, 19–28.

Gamble, A. 1994. *The Free Economy and the Strong State*, 2nd edition. Basingstoke: Macmillan.

Gamble, J. 2005. *Identifying, Assessing and Managing Risk in the Context of Policing Domestic Violence*. London: ACPO.

Gandy, M. 2004. Rethinking urban metabolism: water, space and the modern city. *City*, 8, 363–79.

Gannon, M. and Mihorean, K. 2006. *Criminal Victimization in Canada*. Ottawa: Statistics Canada.

Gardner, G. 2008. Rural community development and governance in *New Labour's Countryside: Rural Policy in Britain since 1997*, edited by Woods, M. Bristol: Policy Press, 169–88.

Garland, D. 1996. The limits of the sovereign state: strategies of crime control in contemporary societies. *British Journal of Criminology*, 36, 445–71.

Garland, D. 2001. *The Culture of Control: Crime and Social Order in Contemporary Society*. Chicago: University of Chicago Press.

Garland, J. and Chakraborti, N. 2004. Another country? Community, belonging and exclusion in rural England, in *Rural Racism*, edited by Chakraborti, N. and Garland, J. Cullompton: Willan, 122–40.

George Street Research Ltd 1999. *Crime and the Farming Community: The Scottish Farm Crime Survey, 1998*. Edinburgh: George Street Research Ltd.

George, V. and Wilding, P. 1992. *Ideology and Social Welfare* 2nd edition. London: Routledge.

Gill, A. 2009. Honour killings and the quest for justice in black and minority ethnic communities in the UK. *Criminal Justice Policy Review*, 20, 475–94.

Gill, A. and Rehman, G. 2004. Empowerment through activism: responding to domestic violence in the South Asian Community. *Gender and Development*, 12, 75–82.

Gill, M. and Mawby, R.I. 1990a. *A Special Constable: A Study of the Police Reserve*. Aldershot: Gower.

Gill, M. and Mawby, R.I. 1990b. *Volunteers in the Criminal Justice System: A Comparative Study of Probation, Police and Victim Support*. Milton Keynes: Open University Press.

Gill, M. and Spriggs, A. 2005. *Assessing the Impact of CCTV. Home Office Research Study no.292*. London: Home Office.

Gilling, D. 1997. *Crime Prevention: Theory, Policy and Politics*. London: UCL Press.

Gilling, D. 2007. *Crime Reduction and Community Safety: New Labour and the Politics of Local Crime Control*. Cullompton: Willan.

Gilling, D. and Pierpoint, H. 1999. Crime prevention in rural areas, in *Crime and Conflict in the Countryside*, edited by Dingwall, G. and Moody, S. Cardiff: University of Wales Press.

Girling, E., Loader, I. and Sparks, R. 2000. *Crime and Social Change in Middle England*. London: Routledge.

Goldstein, H. 1990. *Problem-Oriented Policing*. New York: McGraw-Hill.

Goodwin, M. 1998. The governance of rural areas: some emerging research agendas. *Journal of Rural Studies*, 14, 5–12.

Goodwin, M. and Painter, J. 1996. Local governance, the crises of Fordism and the changing geographies of regulation. *Transactions of the Institute of British Geographers*, 21, 635–48.

Grabb, E. 2004. Economic power in Canada: corporate concentration, foreign ownership, and state involvement, in *Social Inequality in Canada*, edited by Curtis, J. *et al*. Toronto: Pearson Prentice Hall, 20–30.

Grabb, E. and Curtis, J. 2005. *Regions Apart: The Four Societies of Canada and the United States*. Toronto: Oxford University Press.

Grace, S. 1995. *Policing Domestic Violence in the 1990s. Home Office Research Study No. 139*. London: HMSO.

Graham, S. 1999. The eyes have it – CCTV as the fifth utility. *Town and Country Planning*, 68, 312–15.

Graham, S., Brooks, J. and Heery, D. 1996. Towns on television: closed circuit TV in British towns and cities, *Local Government Studies*, 223, 1–27.

Grahame, K. 1926. *The Wind in the Willows*. London: Methuen.

Gray, J. 2000. The common agricultural policy and the re-invention of the rural in the European community. *Sociologia Ruralis*, 40, 30–52.

Guthrie-Smith, H. 1921/1999. *Tutira: The Story of a New Zealand Sheep Station*. Seattle and London: University of Washington Press.

Hadfield, P. 2006. *Bar Wars: Contesting the Night In Contemporary British Cities*. Oxford: Oxford University Press.

Haight, W., Jacobsen, T., Black, J., Kingery, L., Sheridan, K. and Mulder, C. 2005. "In These Bleak Days": Parent methamphetamine abuse and child welfare in the rural midwest. *Children and Youth Services Review*, 27, 949–71.

Halfacree, K. 1996a. Out of place in the country: travellers and the "rural idyll". *Antipode*, 281, 42–72.

Halfacree, K. 1996b. Trespassing against the rural idyll: the Criminal Justice and Public Order Act 1994 and access to the countryside, in *Rights of Way. Policy, Culture and Management*, edited by Watkins, C. London: Pinter, 179–93.

Halfacree, K. 2003. Landscapes of rurality: rural others/other rurals, in *Studying Cultural Landscapes*, edited by Roberston, I. and Richards, P. London: Arnold, 141–69.

Halfacree, K. 2006a. Rural space: constructing a three-fold architecture, in *Handbook of Rural Studies*, edited by Cloke, P. *et al.* London: Sage, 44–62.

Halfacree, K. 2006b. From dropping out to leading on? British counter-cultural back-to-the-land in a changing rurality. *Progress in Human Geography*, 30, 309–36.

Halfacree, K. 2007. Trial by space for a 'radical rural': introducing alternative localities, representations and lives. *Journal of Rural Studies*, 23, 125–141.

Halfacree, K. and Boyle, P. 1998. Migration, rurality and the post-productivist countryside, in *Migration into Rural Areas: Theories and Issues*, edited by Boyle, P. and Halfacree, K. Chichester: Wiley, 1–12.

Halseth, G. and Ryser, L. 2006. Trends in service delivery: examples from rural and small town Canada, 1998 to 2005. *Journal of Rural and Community Development*, 12: 69–90.

Hamer, D. 1988. *The New Zealand Liberals: The Years of Power, 1891–1912*. Auckland: Auckland University Press.

Hanmer, J. and Griffiths, S. 2000. *Reducing Domestic Violence ... What Works? Policing Domestic Violence, Crime Reduction Research Series Briefing Note*. London: Home Office.

Hansard 2006. *Parliamentary Debates Thursday, 11 May 2006 for Inclusion in Volume 631*. Wellington: New Zealand.

Harding, R., Morgan, F., Ferrante, A., Loh, N. and Fernandez, J. 1998. *Rural Crime and Safety in Western Australia. Crime Research Centre Report for the Western Australia Regional Development Council*. Perth: Government of Western Australia, Department of Commerce and Trade.

Harris, R.P. and Worthen, D. 2003. African Americans in rural America, in *Challenges for Rural America in the Twenty-First Century*, edited by Brown, D. *et al.* Pennsylvania: The Pennsylvania State University Press, 32–42.

Hauber, A., Hofstra, B., Toornvliet, L. and Zandbergen, A. 1996. Some new forms of functional social control in the Netherlands and their effects. *British Journal of Criminology*, 36, 199–219.

Hawes, D. and Perez, B. 1995. *The Gypsy and the State*. Bristol: School of Advanced Urban Studies Publications.

Healey, P. 1998. Building institutional capacity through collaborative approaches to urban planning. *Environment and Planning A*, 29, 1531–46.

Hempel, L. and Töpfer, E. 2004. *CCTV in Europe: Final Report, Urban Eye Working Paper No. 15*. [Online: Urban Eye]. Available at: www.urbaneye.net/results/ue_wp15.pdf [Accessed 13th May 2010].

Henderson, S. 1998a. *Drugs Prevention in Rural Areas: Final Report to the Home Office Drugs Prevention Initiative.* London: Home Office.

Henderson, S. 1998b. *Service Provision to Women Experiencing Domestic Violence in Scotland. Crime and Criminal Justice Research Findings No. 20.* Edinburgh: The Scottish Office.

Herbert, S. 1996. The geopolitics of the police: Foucault, disciplinary power and the tactics of the Los Angeles Police Department. *Political Geography* 15, 47–57.

Herbert, S. 2006. *Citizens, Cops and Power: Recognizing the Limits of Community.* Chicago: The University of Chicago Press.

Herbert, S. 2009. Policing in *The Dictionary of Human Geography 5th Edition*, edited by Gregory, D. *et al.* Oxford: Wiley, 544–45.

Herbert-Cheshire, L. 2000. Contemporary strategies for rural community development in Australia: a governmentality perspective. *Journal of Rural Studies*, 16, 203–15.

Herdale, G. and Stelfox, P. 2008. National intelligence model, in *Dictionary of Policing*, edited by Newburn, T. and Neyroud, P. Cullompton: Willan Publishing, 173–75.

Hester, R. 1999. Policing new age travellers: conflict and control in the countryside? In *Crime and Conflict in the Countryside*, edited by Dingwall, G. and Moody, S. Cardiff: University of Wales Press, 130–45.

Hetherington, K. 2000. *New Age Travellers.* London: Cassell.

Higgins, V. and Lockie, S. 2002. Re-discovering the social: neo-liberalism and hybrid practices of governing in rural natural resource management. *Journal of Rural Studies*, 18, 419–28.

Hill, R. 1995. *The Iron Hand in the Velvet Glove: The Modernisation of Policing in New Zealand, 1886–1917.* Palmerston North: Dunmore Press.

HM Government 2008. *Drugs: Protecting Families and Communities. The 2008 Drug Strategy.* London: Home Office.

Hoare, J. 2009. *Drug Misuse Declared: Findings from the 2008/09 British Crime Survey England and Wales.* London: Home Office.

Hobbs, D., Winlow, S., Hadfield, P. and Lister, S. 2005. Violent hypocrisy: governance and the night-time economy. *European Journal of Criminology*, 22, 161–83.

Hoffman, J. 1992. Rural policing. *Law and Order*, 40, 20–24.

Hogg, R. and Carrington, K. 1998. Crime, rurality and community, *Australian and New Zealand Journal of Criminology*, 31, 160–81.

Hogg, R. and Carrington, K. 2003, Violence, spatiality and other rurals, *Australian and New Zealand Journal of Criminology*, 36, 293–319.

Hogg, R. and Carrington, K. 2006. *Policing the Rural Crisis.* Annandale: Federation.

Hoggart, K. 1990. Let's do away with rural, *Journal of Rural Studies*, 6, 245–57.

Holloway, S. 2003. Outsiders in rural society? Constructions of rurality and nature-society relations in the racialisation of English Gypsy-Travellers, 1869–1934. *Environment and Planning D: Society and Space*, 21, 695–715.

Holloway, S. 2004. Rural roots, rural routes: discourse of rural self and travelling other in debates about the future of Appleby New Fair, 1945–1969. *Journal of Rural Studies*, 20, 143–56.

Holloway, S. 2005. Articulating otherness? White rural residents talk about Gypsy-Travellers. *Transactions of the Institute of British Geographers*, 30, 351–367.

Home Office 2004. *National Policing Plan 2005–08. Safer, Stronger Communities.* London: Home Office.

Home Office 2008. *Justice With Safety: Domestic Violence Courts Review 2007–2008.* London: Home Office.

Hope, T. 1995. Building a safer society: strategic approaches to crime prevention, in *Community Crime Prevention: Crime and Justice, A Review of Research, Volume 19*, edited by Tonry, M. and Farrington, D.P. Chicago: University of Chicago Press, 21–89.

Hope, T. 2005. The new local governance of community safety in England and Wales. *Canadian Journal of Criminology and Criminal Justice*, 47, 367–87.

Hope, T. and R. Sparks (eds) 2000. *Crime, Risk and Insecurity: Law and Order in Everyday Life and Political Discourse.* London: Routledge.

Hopkins, N. 2000. Fuel crisis: police opt for softly, softly approach. *The Guardian* 13 September 2000, 2.

Horrall, S.W. 1980. The Royal North-West Mounted Police and labour unrest in western Canada. *Canadian Historical Review*, 16: 169–90.

Hough, M. and Mayhew, P. 1983. *The First British Crime Survey.* London: HMSO.

Hughes, G. 2002. Crime and disorder reduction partnerships, in *Crime Prevention and Community Safety: New Directions*, edited by Hughes, G. *et al.* London: Sage 123–41.

Hughes, G. 2007. *The Politics of Crime and Community.* Basingstoke: Palgrave Macmillan.

Hughes, G., McLaughlin, E. and Muncie, J. (eds) 2002. *Crime Prevention and Community Safety: New Directions.* London: Sage Publications.

Ilbery, B. and Bowler, I. 1998. From agricultural productivism to post-productivism, in *The Geography of Rural Change*, edited by B. Ilbery. Harlow: Longman, 57–84.

IPCC 2006. *Report into the Policing of the Countryside Alliance Pro-Hunting Demonstrations on Wednesday 15 September 2004 at Parliament Square, London.* London: Independent Police Complaints Commission.

Jachmann, H. and Billiouw, M. 1997. Elephant poaching and law enforcement in the central Luangwa Valley, Zambia. *Journal of Applied Ecology*, 34, 233–44.

Jacobson, J. and Saville, E. 1999. *Neighbourhood Warden Schemes: An Overview.* London: Home Office Policing and Reducing Crime Unit.

Jaschke, H.-G., Bjorgo, T., del Barrio Romero, F., Kwanten, C., Mawby, R. and Pagon, M. 2007. *Perspectives on Police Science in Europe*. Bramshill: CEPOL.

James, Z. 2004. *New Travellers, New Policing? Exploring the Policing of New Traveller Communities under the Criminal Justice and Public Order Act 1994*. PhD. University of Surrey.

James, Z. 2005. Eliminating communities? Exploring the implications of policing methods used to manage New Travellers. *International Journal of the Sociology of Law*, 33, 159–68.

James, Z. 2006. Policing space: managing new travellers in England. *British Journal of Criminology*. 46 3: 470–85.

James, Z. 2007a. Policing marginal spaces: controlling gypsies and travellers. *Criminology and Criminal Justice: An International Journal*, 7, 367–89.

James, Z. 2007b. *Enforcing Boundaries? Policing Gypsies and Travellers in a New Europe. European Society of Criminology Conference 2007*. Bologna: Universita Di Bologna.

James, Z. and Browning, R. 2008. *Policing Strangers: A National Study of the Management of Gypsies and Travellers. British Society of Criminology Conference 2008*. Huddersfield: University of Huddersfield.

Jammes, J.R.J. 1982. *The French Gendarmerie*. Bradford: MCB Publications.

Jaschke, H. and Neidhardt, K. 2007. A modern police science as an integrated academic discipline: a contribution to the debate on its fundamentals. *Policing and Society*, 17, 303–20.

Jay, M. 2007. The political economy of a productivist agriculture: New Zealand dairy discourses. *Food Policy*, 32:2, 266–79.

Jay, M., Morad, M. and Bell, A. 2003. Biosecurity, a policy dilemma for New Zealand. *Land Use Policy*, 20:2, 121–9.

Jennett, C. 2001. Policing and Indigenous peoples in Australia, in *Policing the Lucky Country*, edited by Enders, M. and Dupont, B. Leichardt, NSW: Hawkins Press, 50–69.

Jensen, L., McLaughlin, D. and Slack, T. 2003. Rural poverty: the persisting challenge, in *Challenges for Rural America in the 21st Century*, edited by Brown, D. and Swanson, L. Pennsylvania: The State University Press, 118–34.

Jentsch, B. and Shucksmith, M. (eds) 2004. *Young People in Rural Areas of Europe*. Aldershot: Ashgate.

Jiao, A.Y. 2001. Degrees of urbanism and police orientations: testing preferences for different policing approaches across urban, suburban, and rural areas. *Police Quarterly*, 43, 361–87.

Jobes, P.C. 2002. Effective officer and good neighbour: problems and perceptions among police in rural Australia. *Policing*, 25.2, 256–73.

Jobes, P.C. 2003. Human ecology and rural policing: a grounded theoretical analysis of how personal constraints and community characteristics influence strategies of law enforcement in rural New South Wales, Australia. *Police Practice and Research*, 4, 3–19.

Jobes, P., Donnermeyer, J., Barclay, E. and Weinand, H. 2000. *A Qualitative and Quantitative Analysis of the Relationship between Community Cohesiveness and Rural Crime, Part 2*. Armidale: Institute for Rural Futures, University of New England.

Johnson, L. 1996. What is vigilantism? *British Journal of Criminology*, 36.2, 220–36.

Johnston, L. and Shearing, C. 2003. *Governing Security*. London: Routledge.

Johnson, R. and Rhodes, T.N. 2009. Urban and small town comparison of citizen demand for police services. *International Journal of Police Science and Management*, 11, 27–38.

Johnstone, C. 2002 Realising the 'fifth utility'? *Town and Country Planning*, 1st November.

Jones, O. 1997. Little figures, big shadows. Country childhood stories, in *Contested Countryside Cultures. Otherness, Marginalisation and Rurality*, edited by Cloke, P. and Little, J. London: Routledge, 158–79.

Jones, T. 1995. *Policing and Democracy in the Netherlands*. London: Policy Studies Institute.

Jones, T. and Newburn, T. 1995. How big is the private security sector? *Policing and Society*, 5, 221–32.

Jones, T. and Newburn, T. 2002. Learning from Uncle Sam? Exploring U.S. influences on British crime control policy. *Governance*, 15: 97–119.

Jones, T. and Newburn, T. (eds) 2006. *Plural Policing: A Comparative Perspective*. Abingdon: Routledge.

Journal of Rural Studies 18: 2 2002. Special Issue: Young Rural Lives.

Katz, C. 1998. Whose nature, whose culture?: private productions of space and the 'preservation' of nature, in *Remaking Reality. Nature at the Millennium*, edited by Braun, B. and Castree, N. London: Routledge, 46–63.

Kelso, B. 1993. Hunting for Cons. *Africa Report*, July–August, 68–9.

Kenrick, D. and Clark, C. 1999. *Moving On: The Gypsies and Travellers of Britain*. Hatfield: University of Hertfordshire Press.

King, M. 2003. *The Penguin History of New Zealand*. Auckland, Penguin Books.

Koch, B. 1998. *The Politics of Crime Prevention*. Aldershot: Ashgate.

Koskela, H. 2000. 'The gaze without eyes': video surveillance and the changing nature of urban space. *Progress in Human Geography*, 42, 243–65.

Koskela, H. and Pain, R. 2000. Revisiting fear and place: women's fear of attack and the built environment. *Geoforum*, 31, 269–80.

Krimmel, J.T. 1997. Northern York county police consolidation experience: an analysis of the consolidation of police services in eight Pennsylvania rural communities. *Policing*, 20.3, 497–507.

Laffont, H. and Meyer, P. 1990. *Le Nouvel Ordre Gendarmique*. Paris: Le Seuil.

LaFree, G., Bursik, R., Short, J. and Taylor, R. 2000. *Criminal Justice 2000 Series. Volume 1. The Nature of Crime: Continuity and Change.* Washington DC: US Department of Justice, National Institute of Justice.

Lasch, C. 1977. *Haven in a Heartless World: The Family Besieged.* New York: Basic Books.

Lawtey, A. and Deane, M. 2001. *Making Rural Communities Safer: Consultation on Community Safety.* London: NACRO.

Laycock, G. and Tilley, N. 1995. *Policing and Neighbourhood Watch: Strategic Issues.* London: Home Office Crime Prevention and Detection series no. 60.

Le Heron, R. and Pawson, E. 1996. *Changing Places. New Zealand in the Nineties.* Auckland: Longman Paul.

Lee, M. 2007. Fear, law and order and politics: tales of two rural towns, in *Crime in Rural Australia: Integrating Theory, Research and Practice*, edited by Barclay, E., Scott, J., Donnermeyer, J. and Hogg, R. Sydney: Federation press.

Lefebvre, H. 1991 [1974]. *The Production of Space.* Oxford: Blackwell.

Leiderbach, J. and Frank, J. 2003. Policing Mayberry: the work routines of small-town and rural officers. *American Journal of Criminal Justice*, 28.1, 53–72.

Lemieux, A. and Clarke, R. 2009. The international ban on ivory sales and its effect on elephant poaching in Africa. *British Journal of Criminology*, 49, 451–71.

Levison, D. and Harwin, N. 2000. *Reducing Domestic Violence...What Works? Accommodation Provision.* London: Home Office.

Lewis, N., Larner, W. and Le Heron, R. 2008. The New Zealand designer fashion industry: making industries and co-constituting political projects. *Transactions of the Institute of British Geographers*, NS, 33, 42–59.

Lewis, S.H. 2003. *Unspoken Crimes: Sexual Assault in Rural America.* Enola, PA: National Sexual Violence Resource Center.

Leyshon, M. 2002. On being "in the field": practice, progress and problems in research with young people in rural areas. *Journal of Rural Studies*, 18, 179–91.

Liepins, R. 2000. New energies for an old idea: reworking approaches to "community" in contemporary rural studies. *Journal of Rural Studies*, 16, 23–35.

Lignereux, A. 2002. *Gendarmes et Policiers dans la France de Napoléon.* Paris: La Documentation Français.

Lipsky, M. 1980. *Street-level Bureaucracy: Dilemmas of the Individual in Public Services.* New York: Basic books.

Lithopoulos, S. and Rigakos, G.S. 2005. Neo-liberalism, community and police regionalism in Canada: A critical empirical analysis. *Policing*, 28, 337–52.

Little, J. 1999. Otherness, representation and the cultural construction of rurality. *Progress in Human Geography*, 23, 437–42.

Little, J. and Panelli, R. 2003. Gender research in rural geography. *Gender, Place and Culture*, 10, 281–89.

Little, J., Panelli, R. and Kraack, A. 2005. Women's fear of crime: a rural perspective. *Journal of Rural Studies*, 21, 151–63.

Lloyd, L. 1993. Proposed reform of the 1968 Caravan Sites Act: producing a problem to suit a solution? *Critical Social Policy*, 38, 77–85.

Loveday, B. 1999. Government and accountability of the police, in *Policing Across the World: Issues for the Twenty-first Century*, edited by Mawby, R.I. London: UCL Press, 132–50.

Loveday, B. and Reid, A. 2003. *Going Local: Who Should Run Britain's Police*. London: Policy Exchange.

Lowe, P., Murdoch, J., Marsden, T.K., Munton, R. and Flynn, A. 1993. Regulating the new rural spaces: the uneven development of land. *Journal of Rural Studies*, 9, 205–22.

Lowe, R. and Shaw, W. 1993. *Voices of the New Age Nomads*. London: Fourth Estate.

Luc, J. 2002. Le gendarme, un soldat qui a pris racine? in *Gendarmerie, Etat et Société au XIXe Siècle*, edited by Luc, J.-N. Paris: Publications de la Sorbonne, 315–43.

Luc, J. (ed.) 2005. *Histoire de la Maréchaussée et de la Gendarmerie. Guide de Recherché*. Maisons-Alfort, Service historique de la Gendarmerie Nationale.

Lukes, S. 1976 *Power: A Radical View*. London: Macmillan.

Lynch, M. 1990. The greening of criminology: a perspective for the 1990s. *The Critical Criminologist*, 2, 11–12.

Lyson, T.A. and Tolbert, C.M. 2003. Civil society, civic communities, and rural development, in *Challenges for Rural America in the 21st Century*, edited by Brown, D. and Swanson, L. Pennsylvania: The State University Press, 228–40.

Mackay, A. 2000. *Reaching Out: Women's Aid in a Rural Area*. East Fife. Scotland: Women's Aid.

Maguire, M. and John, T. 2006. Intelligence-led policing, managerialism and community engagement: competing priorities and the role of the National Intelligence Model in the UK. *Policing and Society*, 16, 67–85.

Maguire, B., Faulkner, W., Mathers, R., Rowland, C. and Wozniak, J. 1991. Rural police job functions. *Police Studies*, 14, 180–87.

Manning, P. 1977. *Police Work: The Social Organisation of Policing*. Cambridge, MA: MIT Press.

Martin, E. and Milliken, T. 2005. *No Oasis: The Egyptian Ivory Trade in 2005*. East/Southern Africa Cambridge, UK: TRAFFIC.

Martin, G. 2002. New Age Travellers: uproarious or uprooted? *Sociology*, 36, 723–35.

Matelly, J. 2000, *Gendarmerie et Crimes de Sang*. Paris: l'Harmattan.

Matelly, J. and Mouhanna, C. 2007. *Police, des Chiffres et des Doutes* Paris: Michalon.

Matthews, H., Taylor, M., Sherwood, K., Tucker, F. and Limb, M. 2000. Growing-up in the countryside: children and the rural idyll, *Journal of Rural Studies*, 16, 141–53.

Mawby, R.C. and Wright, A. 2008. The police organisation, in *Handbook of Policing* Second Edition, edited by Newburn, T. Cullompton: Willan Publishing, 224–52.

Mawby, R.I. 1990. *Comparative Policing Issues: The British and American Experience in International Perspective.* London: Routledge.

Mawby, R.I. (ed.) 1999. *Policing Across the World: Issues for the Twenty-first Century.* London: UCL Press.

Mawby, R.I. 2004. Myth and reality in rural policing: perceptions of police in a rural county of England. *Policing*, 27, 431–46.

Mawby, R.I. 2007. Crime, place and explaining rural hotspots. *International Journal of Rural Crime*, 1, 21–43.

Mawby, R.I. 2009. Perceptions of police and policing in a rural county of England. *International Journal of Police Science and Management*, 11, 39–53.

Mayall, D. 1988. *Gypsy-Travellers in Nineteenth Century Society.* Cambridge: Cambridge University Press.

Mayall, D. 2004. *Gypsy Identities 1500–2000: From Egipcyans and Moon-Men to the Ethnic Romany.* London: Routledge.

Mazzerole, L., Marchett, E. and Lindsay, A. 2003. Policing and the plight of indigenous Australians: past conflicts and present challenges. *Police and Society*, 7, 75–102.

McCarry, M. and Williamson, E. 2009. *Violence Against Women in Rural and Urban Areas.* Bristol: University of Bristol.

McGrath, B. 2001. "A problem of resources". Defining youth encounters in education, work and housing. *Journal of Rural Studies*, 17, 481–95.

McKay, G. 1996. *Senseless Acts of Beauty.* London: Verso.

McMullan, J. 1987. Crime, laws and order in early-modern England. *British Journal of Criminology*, 27, 252–74.

Mears, D.P., Scott, M.L. and Bhati, A. 2007a. A process and outcome evaluation of an agricultural crime prevention initiative. *Criminal Justice Policy Review*, 18, 51–80.

Mears, D., Scott, M. and Bhati, A. 2007b. Opportunity theory and agricultural crime victimization. *Rural Sociology*, 72:2, 151–84.

Mears, D.P., Scott, M., Bhati, A., Roman, J., Chalfin, A. and Jannetta, J. 2007c. *A Process and Impact Evaluation of the Agricultural Crime, Technology, Information, and Operations Network ACTION Program.* Washington DC: The Urban Institute.

Merrick 1996. *Battle for the Trees.* Leeds: Godhaven Ink.

Messer, K. 2000. The poacher's dilemma: the economics of poaching and enforcement. *Endangered Species Update*, 17, 50–56.

Millie, A. (ed.) 2009. *Securing Respect: Behavioural Expectations and Anti-Social Behaviour.* Bristol: The Policy Press.

Miller, W.R. 1977. *Cops and Bobbies: Police Authority in New York and London, 1830–1870.* Chicago: University of Chicago Press.

Milliken, T. 2004. *Domestic Ivory Markets: Where They Are and How They Work.* Cambridge, UK: TRAFFIC.

Milliken, T., Burn, R. and Sangalakula, L. 2007. *The Elephant Trade Information System ETIS and the Illicit Trade in Ivory: A Report to the 14th Meeting of the Conference of Parties to CITES.* Cambridge, UK: TRAFFIC.

Milliken, T., Pole, A. and Huongo, A. 2006. *No Peace for Elephants: Unregulated Domestic Ivory Markets in Angola and Mozambique.* Cambridge, UK: TRAFFIC.

Mingay, G. (ed.) 1989. *The Unquiet Countryside.* London: Routledge.

Ministry for Agriculture and Forestry 2008. *Briefing for Incoming Ministers.* Wellington: Ministry for Agriculture and Forestry.

Mirrlees-Black, C. and Byron, C. 1994. S*pecial Considerations Issues for the Management and Organisation of the Volunteer Police.* London: Home Office.

Moats, J.B. 2007. *Agroterrorism: A Guide for First Responders.* College Station, Texas: Texas AandM University Press.

Moley, S. 2008. Public perceptions, in *Crime in England and Wales: Findings from the British Crime Survey and Police Recorded Crime*, edited by Kershaw, C. *et al.* London: Home Office, 117–42.

Monjardet, D. 1996. *Ce que Fait la Police, Sociologie de la Force Publique* Paris: La Découverte.

Monkkonen, E. 1981. *Police in Urban America, 1860–1920.* Cambridge: Cambridge University Press.

Moody, S. 1999. Rural neglect: the case against criminology, in *Crime and Conflict in the Countryside*, edited by Dingwall, G. and Moody, S. Cardiff: University of Wales Press, 8–29.

Moore, D. 1992. Policing rural communities; a northern territory police perspective, in *AIC Conference Proceedings; No. 5*, edited by Vernon, J. and McKillop, S. Canberra: Australian Institute of Criminology, 199–126 Available at: http://www.aic.gov.au/publications/proceedings/05/dave-moore.html [Accessed June 2008].

Morley, R. and Mullender, A. 1994. *Preventing Domestic Violence to Women. Police Research Group Crime Prevention Series Paper 48.* London: Home Office.

Mormont, M. 1990 Who is rural? Or, how to be rural: towards a sociology of the rural, in *Rural Restructuring: Global Processes and their Local Responses*, edited by Marsden, T., Lowe, P. and Whatmore, S. London: Fulton, 21–44.

Morris, R. and Clements, L. 2002. *At What Cost? The Economics of Gypsy and Traveller Encampments.* Bristol: The Policy Press.

Morrison, W.R. 1985. *Showing the Flag: The Mounted Police and Canadian Sovereignty in the North, 1894–1925.* Vancouver: University of British Columbia Press.

Mouhanna, C. 1997. *Logique de Proximité et Exigence Gestionnaire: où va la Gendarmerie Départementale?* Paris: Centre de Sociologie des Organisations-Centre D'etudes en Sciences Sociales de la Défense.

Mouhanna, C. 2000. *Quel Service pour quel Public? Une Tentative D'évaluation Chiffrée de L'image de la Police dans la Population Face à La Territorialisation.* Paris: CSO-IHESI.

Mouhanna, C. 2001. Faire le gendarme: de la souplesse informelle à la rigueur bureaucratique. *Revue Française de Sociologie*, 42, 31–55.

Mouhanna, C. 2002a. Une police de proximité judiciarisée. *Déviance et Sociétél*, 22, 163–82.

Mouhanna, C. 2002b. *Le Gendarme et la Règle, Centre de Sociologie des Organisations.* Paris: Centre de prospective de la Gendarmerie Nationale.

Mucchielli, L. 2007. *Gendarmes et Voleurs, de L'Évolution de la Délinquance aux Défis du Metier.* Paris: L'harmattan.

Muchembled, R. 2008. *Une Histoire de la Violence – De la Fin du Moyen-Age à nos Jours.* Paris: Seuil L'Univers historique.

Muncie, J. 2004. *Youth and Crime.* London: Sage.

Murdoch, J. and Pratt, A. 1993. Rural studies: modernism, postmodernism and the 'post-rural'. *Journal of Rural Studies*, 9, 411–27.

Murdoch, J., Lowe, P., Ward, N. and Marsden, T. 2003. *The Differentiated Countryside.* London: Routledge.

Murdoch, J. and Pratt, A. 1997. From the power of topography to the topography of power. A discourse on strange ruralities, in *Contested Countryside Cultures*, edited by Cloke, P. and Little, J. London: Routledge, 51–69.

Murdoch, J. and Ward, N. 1997. Governmentality and territoriality. The statistical manufacture of Britain's 'national farm'. *Political Geography*, 16, 307–324.

National Policing Improvement Agency 2008. *Neighbourhood Policing in Rural Communities.* [Online: National Policing Improvement Agency]. Available at: www.npia.police.uk [Accessed 15th May 2010].

National Strategy for Neighbourhood Renewal 2000. *Neighbourhood Wardens.* London: Cabinet Office, Social Exclusion Unit.

Naylor, E.L. 1994. Unionism, peasant protest and the reform of French agriculture. *Journal of Rural Studies*, 10, 263–73.

Neal, D. 1991. *The Rule of Law in a Penal Colony: Law and Power in Early New South Wales.* Cambridge: Cambridge University Press.

New South Wales Crime Prevention Division NSWCPD 2008. *CPD Projects.* [Online: New South Wales Crime Prevention Division]. Available at: http://www.lawlink.nsw.gov.au/lawlink/cpd/ll_cpd.nsf/pages/CPD_projects [Accessed June 2008].

New South Wales Police 2008. *About Us.* [Online: New South Wales Police]. Available at: http://www.police.nsw.gov.au/about_us [Accessed June 2008].

New York Times Magazine 2001. A Watchful State, 7th October.

New Zealand Police 2006. *Police Act Review Issues Paper 4: Community Engagement.* Wellington: New Zealand Police.

Newburn, T. 2002. Atlantic crossings: "policy transfer" and crime control in the USA and Britain. *Punishment and Society*, 4, 165–94.

Newburn, T. 2003. The future of policing, in *Handbook of Policing*, edited by Newburn, T. Cullompton: Willan, 707–21.

Newburn, T. (ed.) 2003. *Handbook of Policing*. Cullompton: Willan Publishing.

Newby, H., Bell, C., Rose, D. and Saunders, P. 1978. *Property, Paternalism and Power*. London: Hutchinson.

Newsbeat 2007a. *Country Policing – The True Story*. [Online: Western Australia Police Newsbeat, Issue 42, June]. Available at: http://www.police.wa.gov.au/Portals/11/PDFs/Newsbeat_42.pdf [Accessed June 2008].

Newsbeat 2007b. *Merredin – It's What You Make It*. [Online: Western Australia Police Newsbeat, Issue 43, September]. Available at: http://www.police.wa.gov.au/Portals/11/PDFs/Newsbeat_43.pdf [Accessed June 2008].

Neyroud, P. 2004. Why the police need voluntary helpers. *New Statesman*, Special Supplement 17th May 2004, 28.

ni Shuinéar, S. 1997. Why do Gaujos hate Gypsies so much, anyway? A case study, in *Gypsy Politics and Traveller Identity*, edited by Acton, T. Hatfield: University of Hertfordshire Press, 26–53.

Niner, P. 2003. *Local Authority Gypsy/Traveller Sites in England*. London: Office of the Deputy Prime Minister.

Niner, P. 2004a. Accommodating nomadism? An examination of accommodation options for Gypsies and Travellers in England. *Housing Studies*, 19, 141–159.

Niner, P. 2004b. *Counting Gypsies and Travellers: A Review of the Gypsy Caravan Count System*. London: Office of the Deputy Prime Minister.

Niner, P. 2007. *Preparing Regional Spatial Strategy Reviews on Gypsy and Travellers by Regional Planning Bodies*. London: DCLG.

Norfolk Eastern Daily Press 2008. Protests over travellers site. [Online Norfolk Eastern Daily Press, 5th September]. Available at: http://new.edp24.co.uk/content/news/story.aspx?brand=EDPOnlineandcategory=NewsandtBrand=edponlineandtCategory=newsanditemid=NOED05%20Sep%202008%2011%3A55%3A59%3A133 [Accessed December 2008].

Norris, C. and Armstrong, G. 1999. *The Maximum Surveillance Society: The Rise of CCTV*. Oxford: Berg.

North Yorkshire Police 2008. *ACPO Hunting Act Conference, Friday, October 10, 2008, Report by North Yorkshire Police*. [Online: North Yorkshire Police]. Available at: www.policeoracle.com [Accessed 12 December 2008].

Northants Evening Telegraph 2008. Travellers could get an eco village. [Online: Northants Evening Telegraph, 15th September]. Available at: http://www.northantset.co.uk/news/Travellers-could-get-an-eco.4490421.jp [Accessed April 2009].

NZPD 1871. Diseased cattle bill. *New Zealand Parliamentary Debates*, 11, 153–56.

NZPD 1896. Noxious weeds act. *New Zealand Parliamentary Debates*, 93, 156.

NZPD 1898. Dairy industry bill. *New Zealand Parliamentary Debates*, 105, 60.

NZPD 1998. Ministry of Agriculture and Forestry Restructuring Bill. *New Zealand Parliamentary Debates*, 570, 10937–8.

NZPD 2007. State-Owned Enterprises AgriQuality Limited and Asure New Zealand Limited Bill. *New Zealand Parliamentary Debates*, 638, 8558–74.

Office of Crime Prevention 2004. *Preventing Crime. State Community Safety and Crime Prevention Strategy.* Perth, WA: Office of Crime Prevention.

Office of Crime Prevention 2005. *WA Rural Crime Prevention Strategy: Consultation Draft.* Perth, WA: Office of Crime Prevention, Perth.

O'Connor, M. and Gray, D. 1989. *Crime in a Rural Community.* Sydney: The Federation Press.

ODPM 2006. *Gypsy and Traveller Accommodation Assessments: Draft Practice Guidance.* ODPM: Gypsy and Traveller Unit.

Office of the Deputy Prime Minister 2004. *Neighbourhood Wardens Scheme Evaluation.* Wetherby: Neighbourhood Renewal Unit.

Okely, J. 1983. *The Traveller-Gypsies.* Cambridge: Cambridge University Press.

Ontario Provincial Police 2006. *Rural and Agricultural Crime Team.* [Online: Ontario Provincial Police]. Available at: http://www.opp.ca/Community/CrimePrevention/opp_000282.html [Accessed December 2009].

Orr-Munro, T. 2005. Hunting ban. *Police*, March, 15–17.

Ostrom, E., Parks, R.B. and Whitaker, G.P. 1978. *Patterns of Metropolitan Policing.* Cambridge, Mass: Ballinger.

Panelli, R., Little, J. and Kraack, A. 2004. A community issue? Rural women's feelings of safety and fear in New Zealand. *Gender, Place and Culture*, 11, 445–56.

Parker, H., Williams, L. and Aldridge, J. 2002. The normalization of "sensible" recreational drug use: further evidence from the North West England longitudinal study. *Sociology*, 36, 941–64.

Payne, B.K., Berg, B. and Sun, Y. 2005. Policing in small town America: dogs, drunks, disorder, and dysfunction. *Journal of Criminal Justice*, 33: 31–41.

Peale, K.O. 1990. *Crime and Vandalism on Farms in Kentucky.* Frankfort: Kentucky State University.

Pennings, J. 1999. Crime in rural NSW: a police perspective. Paper to *Conference on Crime in Rural Communities: The Impacts, the Causes, the Prevention*, Armidale, November-December.

Petrow, S. 2001. After Arthur: policing in Van Diemen's land 1837–46, in *Policing the Lucky Country*, edited by Enders, M. and Dupont, B. Leichardt, NSW: Hawkins Press, 176–98.

Phillips, C. 2002. From voluntary to statutory status: reflecting on the experience of three partnerships established under the Crime and Disorder Act 1998, in *Crime Prevention and Community Safety: New Directions*, edited by Hughes, G. *et al.* London: Sage, 163–82.

Philo, C. 1992. Neglected rural geographies: a review. *Journal of Rural Studies*, 8, 193–207.

Pierce, R. 2001. One cop shop: policing in a one man department. *Law and Order*, 49, 107–10.

Police Professional 2009. Moving communities. *Police Professional*. 5th February, 18–21.

Powell, R. 2008. Understanding the stigmatization of Gypsies: power and the dialectics of disidentification. *Housing, Theory and Society*, 25, 87–109.

Power, C. 2004. *Room to Roam: England's Irish Travellers*. London: Action Group for Irish Youth.

Queensland Police 2008. *Stock and Rural Crime Investigation Squads*. [Online: Queensland Police]. Available at: http://www.police.qld.gov.au/Resources/Internet/join/documents/SARCIS.pdf [Accessed June 2008].

Quinton, P. and Morris, J. 2008. *Neighbourhood Policing: The Impact of Piloting and Early National Implementation*. [Online: Home Office Online Report 01/08]. Available at: http://www.homeoffice.gov.uk/rds/pdfs08/rdsolr0108.pdf [Accessed 14th May 2010].

Rai, D.K. and Thiara, R.K. 1997. *Redefining Spaces: The Needs of Black Women and Children using Refuges in England. Their Feelings, Needs and Priority Areas For Development in Refuges*. Bristol: Women's Aid Federation of England.

Reaves, B.A. 2007. *Census of State and Local Law Enforcement Agencies, 2004*. [Online: Washington, DC: US Department of Justice, Bureau of Justice Statistics Bulletin]. Available at: www.ojp.usdoj.gov/bjs/pub/pdf/csllea04.pdf. [Accessed 2nd July 2009].

Reed, M. 2004. The mobilisation of rural identities and the failure of the rural protest movement in the UK, 1996–2001. *Space and Polity*, 8, 25–42.

Reichel, P.L. 1988. Southern slave patrols as a transitional police type. *American Journal of Police*, 7, 52–77.

Reiner, R. 2000. *The Politics of the Police, 3rd edition*. Oxford: Oxford University Press.

Reiner, R. 2002. The Organization and Accountability of the Police, in *The Handbook of the Criminal Justice*, edited by McConville, M. *et al.* Oxford: Oxford University Press, 21–42.

Rephann, T.J. 1999. Links between rural development and crime. *Papers in Regional Science*, 78, 365–86.

Reppetto, T.A. 1978. *The Blue Parade*. New York, NY: Free Press.

Restoule, B.M. 2009. Aboriginal women in the criminal justice system, in *Women and the Criminal Justice System: A Canadian Perspective*, edited by Barker, J. Toronto: Emond Montgomery.

Richards, J. 2001. Moreton Telegraph Station 1902: the native police on Cape York Peninsula, in *Policing the Lucky Country*, edited by Enders, M. and Dupont, B. Leichardt, NSW: Hawkins Press, 96–106.

Richards, L. 2003. *Findings from the Multi-Agency Domestic Violence Murder Reviews in London*. London: Metropolitan Police.

Richardson, J. 2005. *Policing, Gypsies and Travellers: Housing Studies Association Conference 2005*. Lincoln: University of Lincoln.

Richardson, J. 2006. Talking about Gypsies: the notion of discourse as control, *Housing Studies*, 21, 77–96.

Richardson, L. 1992. Parties and political change, in *The Oxford History of New Zealand*, Second Edition, edited by Rice, G. Auckland: Oxford University Press, 201–29.

Robert, M. 1994. *Gamekeepers for the Nation: The Story of New Zealand's Acclimatisation Societies*. Christchurch: Canterbury University Press.

Roberts, M., Cook, D., Jones, P. and Lowther, J. 2001. *Wildlife Crime in the UK: Towards a National Wildlife Crime Unit*. London: Department for Environment, Food and Rural Affairs.

Rose, N. 1999. *Powers of Freedom: Reframing Political Thought*. Cambridge: Cambridge University Press.

Rosenbaum, D. 1988. A critical eye on Neighborhood Watch: does it reduce crime and fear?, in *Communities and Crime Reduction*, edited by Hope, T. and Shaw, M. London: HMSO, 126–45.

Rowe, M. 2004. *Policing, Race and Racism*. Collumpton: Willan.

Royal Canadian Mounted Police 2008a. *Organization of the RCMP*. [Online RCMP]. Available at: http://www.rcmp-grc.gc.ca/about/organi_e.htm [Accessed January 2010].

Royal Canadian Mounted Police 2008b. *Serving Canada's Aboriginal Peoples*. [Online: RCMP]. Available at: http://www.rcmp-grc.gc.ca/aboriginal/aborig_e.htm [Accessed January 2010].

Royal Canadian Mounted Police 2008c. *The RCMP's History*. [Online: RCMP]. Available at: http://www.rcmp-grc.gc.ca/factsheets/pdfs/history_e.pdf [Accessed January 2010].

Rutherford, A. 2002a. *Growing Out of Crime. Second Edition*. Winchester: Waterside Press.

Rutherford, A. 2002b. *Review of Criminal Justice Policy in Jersey*. [Online: St Helier Home Affairs Committee]. Available at: homeaffairs.gov.je/uploads/1227-4070.pdf [Accessed 1st October 2009].

Said, M., Chunge, R., Craig, G., Touless, C., Barnes, R. and Dublin, H. 1995. *African Elephant Database 1995*. Glad, Switzerland: African Elephant Specialist Group, IUCN Species Survival Commission.

Scott, J., Hogg, R., Barclay, E. and Donnermeyer, J. 2007. Introduction, in *Crime in Rural Australia*, edited by Barclay, E. *et al*. Sydney: Federation Press 1–14.

Scott, J. and Jobes, P.C. 2007. Policing in rural Australia: the country cop as law enforcer and local resident, in *Crime in Rural Australia*, edited by Barclay, E. *et al*. Sydney: Federation Press, 127–37.

Scottish Government 2008. *Natural Justice: A Joint Thematic Inspection of the Arrangement in Scotland for Preventing, Investigating and Prosecuting Wildlife Crime*. Edinburgh: Scottish Government.

Scottish Parliament 2007. *Official Report 4 October 2007.* Edinburgh: Scottish Parliament cols. 2493–2538.

Sev'er, A. 2002. *Fleeing the House of Horrors: Women Who Have Left Abusive Partners.* Toronto: University of Toronto Press.

Sharpe, J. 1988. The history of crime in England c.1300–1914. *British Journal of Criminology*, 28, 124–37.

Shepherd, C. 2002. *The Trade of Elephants and Elephant Products in Myanmar.* Cambridge, UK: TRAFFIC.

Sherman, J. 2005. *Men Without Sawmills: Masculinity, Rural Poverty and Family Stability.* Columbia, Missouri: Rural Poverty Research Center.

Short, E. and Ditton, J. 1996. *Does CCTV Prevent Crime? An Evaluation of the Use of CCTV Surveillance Cameras in Airdrie Town Centre.* Edinburgh: Scottish Office Central Research Unit.

Short, E. and Ditton, J. 1998. Seen and now heard: talking to the targets of open-street CCTV. *British Journal of Criminology*, 383, 404–28.

Shucksmith, M. 2004. Young people and social exclusion in rural areas. *Sociologia Ruralis*, 44, 43–59.

Sibley, D. 1981. *Outsiders in Urban Society.* Oxford: Blackwell.

Sibley, D. 1992. Outsiders in society and space, in *Inventing Places. Studies in Cultural Geography*, edited by Anderson, K. and Gale, F. Melbourne: Longman Cheshire, 107–22.

Sibley, D. 1995. *Geographies of Exclusion.* London: Routledge.

Sibley, D. 1997. Endangering the sacred: nomads, youth culture and the English countryside, in *Contested Countryside Cultures*, edited by Cloke, P. and Little, J. London: Routledge, 218–31.

Sibley, D. 2003. Psychogeographies of rural space and practices of exclusion, in *Country Visions*, edited by Cloke, P. Harlow: Pearson, 218–30.

Simpson, J. 2000. *Star-spangled Canadians: Canadians Living the American Dream.* Toronto: HarperCollins.

Sims, L. 2001. *Neighbourhood Watch: Findings from the 2000 British Crime Survey.* London: Home Office.

Sims, L. 2003. Policing and the public, in *Crime in England and Wales 2001/2002: Supplementary Volume*, edited by Flood-Page, C. and Taylor, J. London: HMSO, 105–18.

Sims, V.H. 1988. *Small Town and Rural Police.* Springfield, IL: Charles C. Thomas.

Singer, L. 2004. *Community Support Officer Detention Power Pilot: Evaluation Results.* [Online: Home Office]. Available at: www.crimereduction.gov.uk/policing04.htm [Accessed 5th November 2009].

Smith, K. 2008. *Working Hard for the Money: Trends in Women's Employment, 1970 to 2007 Reports on Rural America 1:5.* Durham, New Hampshire: University of New Hampshire, Carsey Institute.

Smith, K., Barrett, C. and Box, P. 2001. Not necessarily in the same boat: heterogeneous risk assessment among East African pastoralists. *The Journal of Development Studies*, 37, 1–30.

Solicitor General Canada 1996. *Partners in Policing: The Royal Canadian Mounted Police Contract Policing Program.* Ottawa: Ministry of the Solicitor General Canada.

Solicitor General Canada 2004. *First Nations Policing Policy.* Ottawa: Ministry of the Solicitor General Canada.

South, N. and Beirne, P. (eds) 2006. *Green Criminology.* Aldershot: Ashgate.

Southern, R. and James, Z. 2006. *Devon-wide Gypsy and Traveller Housing Needs Assessment.* Plymouth: Social Research and Regeneration Unit, University of Plymouth.

Southgate, P., Bucke, T. and Byron, C. 1995. Parish special constables. *Policing*, 11, 185–93.

Squires, P. 2006. New Labour and the politics of antisocial behaviour, *Critical Social Policy*, 261, 144–68.

Squires, P. 2008. The politics of anti-social behaviour, *British Politics*, 3, 300–23.

Star, P. and Brooking, T. 2007. The Department of Agriculture and Pasture Improvement, *New Zealand Geographer*, 63, 192–201.

States of Jersey 1996. *Report of the Independent Review Body on Police Services in Jersey Clothier Report.* St Helier, Jersey: States of Jersey.

States of Jersey Police 2005. *Annual Report.* St Helier, Jersey: States of Jersey.

Statistics Canada. 2007. *A Comparison of Urban and Rural Crime Rates*. Ottawa: Author.

Stead, P.J. 1983. *The Police of France.* London: Macmillan.

Steden, R. van and Huberts, L. 2006. The Netherlands, in *Plural Policing: A Comparative Perspective*, edited by Jones, T. and Newburn, T. Abingdon: Routledge, 12–33.

Stenson, K. and Watt, P. 1999. Crime, risk and governance in a southern English village, in *Crime and Conflict in the Countryside*, edited by Dingwall, G. and Moody, S. Cardiff: University of Wales Press, 76–93.

Stiles, D. 2004. The ivory trade and elephant conservation. *Environmental Conservation*, 31, 309–21.

Stockdale, J.E. 2002. *Wardens – International Research. Paper to Neighbourhood Renewal Unit Conference on Wardens and Liveability – The International Experience.* London: BAFTA.

Stokes, E. 1996. *Hunting and Hunt Saboteurs: A Censure Study.* [Online: University of East London School of Law, Research Publication Series]. Available at: http://www.uel.ac.uk/law/staff/esarchive.htm [Accessed 17 February 2009].

Storey, D. 2010. Partnerships, people and place: lauding the local in rural development, in *The Next Rural Economies: Constructing Rural Place in a Global Economy*, edited by Halseth, G. *et al.* Oxford: CABI, 155–65.

Storey, D. and Palmer, C. 2002. *Leominster Young Persons Drugs Outreach Project: An Evaluation*. Worcester: Centre for Rural Research, University of Worcester.

Sudgen, G. 1999. Farm crime: out of sight, out of mind: a study of crime on farms in the county of Rutland, England. *Crime Prevention and Community Safety*, 1, 29–36.

Swanson, C.R., Territo, L. and Taylor, R.W. 1988. *Police Administration*, New York: Macmillan.

Sweatman, B. and Cross, A. 1989. The police in the United States, *CJ International*, 5, 11–18.

Taylor, I. and Young, J. 1973. *The New Criminology: For a Social Theory of Deviance*. London: Routledge and Kegan Paul.

Terrill, R.J. 2007. *World Criminal Justice Systems*. Cincinnati: LexisNexis.

The Economist 1998. Under threat: South Africa, *The Economist*, 348 October 10, 502.

The Guardian 2008. Record penalty for grouse moor where poison was found near birds of prey, *The Guardian*, 22nd September 2008.

The Herald 2009. Play for time until the Gypsy Law is ditched. *The Herald*, 29th June 2009.

Thomas, R.H. 1983. *The Politics of Hunting*. Aldershot: Gower.

Thurman, Q.C. and McGarrell, E.F. (eds) 1997. *Community Policing in a Rural Setting*. Cincinnati, OH: Anderson.

Tong, R. 1998. *Feminist Thought: A More Comprehensive Introduction*. Oxford: Westview Press.

Torpey, J. 2000. *The Invention of the Passport: Surveillance, Citizenship and the State*. Cambridge: Cambridge University Press.

Turner, R. 2000. Gypsies and Politics in Britain, *Political Quarterly*, 71, 68–77.

Urry, J. 2007. *Mobilities*. Cambridge: Polity.

US Department of Justice, Bureau of Justice Statistics 2003. *Tribal Law Enforcement, 2000*. Washington DC: US Department of Justice.

US Department of Justice, Bureau of Justice Statistics 2006. *Local Police Departments, 2003*. Washington DC: US Department of Justice.

US Department of Justice, Bureau of Justice Statistics 2007. *Census of State and Local Law Enforcement Agencies, 2004*. Washington DC: US Department of Justice.

Valentine, G. 1989. The geography of women's fear. *Area*, 21, 385–90.

Van Aarde, R. and Jackson, T. 2006. Elephants in Africa. *Africa Geographic*, April, 28–29.

Vanderbeck, R. 2003. Youth, racism, and place in the Tony Martin affair. *Antipode*, 25, 363–84.

Waddington, P. 2000. Public order policing: citizenship and moral ambiguity, in *Core Issues in Policing*, second edition, edited by Leishman, F. *et al*. Harlow: Longman, 156–75.

Walby, K. 2005. Open-street camera surveillance and governance in Canada. *Canadian Journal of Criminology and Criminal Justice*, 474, 655–83.

Walby, S. and Allen, J. 2004. *Domestic Violence, Sexual Assault and Stalking. Findings from the British Crime Survey. Home Office Research Study 276.* London: Home Office.

Walker, S. 1983. *The Police in America: An Introduction.* New York, NY: McGraw-Hill.

Wall, D. 1998. *The Chief Constables of England and Wales: The Socio-legal History of a Criminal Justice Elite.* Aldershot: Ashgate.

Wall, D. 1999. *Earth First! and the Anti-Roads Movement: Radical Environmentalism and Comparative Social Movements.* London: Routledge.

Walsh, P. and White, L. 1999. What it will take to monitor Forest Elephant populations. *Conservation Biology*, 13, 1194–202.

Walters, R. 2004. Criminology and genetically modified food. *British Journal of Criminology*, 44, 151–67.

Walters, R. 2006. Crime, agriculture and the exploitation of hunger. *British Journal of Criminology*, 46, 26–45.

Walton, E. 2002. Liberty and livelihood. *The Field*, August, 50–52.

Wasser, S., Mailand, C., Booth, R., Mutayoba, B., Kisamo, E., Clark, B. and Stephens, M. 2007. Using DNA to track the origin of the largest ivory seizure since the 1989 trade ban. *Proceedings of the National Academy of Sciences*, 104, 4228–233.

Websdale, N. 1995a. An ethnographic assessment of the policing of domestic violence in rural eastern Kentucky. *Social Justice*, 22, 102–22.

Websdale, N. 1995b. Rural woman abuse: the voices of Kentucky women. *Violence Against Women*, 1, 309–38.

Websdale, N. 1998. *Rural Woman Battering and the Justice System: An Ethnography.* Thousand Oaks, California: Sage.

Webster, W. 2009. CCTV policy in the UK: reconsidering the evidence base. *Surveillance and Society*, 61, 10–22.

Weisheit, R. 2008. Making Methamphetamine. *Southern Rural Sociology*, 23, 78–107.

Weisheit, R., Falcone, D. and Wells, L.E. 1996. *Crime and Policing in Rural and Small-town Rural America.* Prospect Heights, IL: Waveland Press.

Weisheit, R.A., Falcone, D.N. and Wells, L.E. 2006. *Crime and Policing in Rural and Small-Town America*, Third Edition. Prospect Heights, IL: Waveland Press.

Weisheit, R.A., Wells, L.E. and Falcone, D.N. 1994. Community policing in small town and rural America. *Crime and Delinquency*, 40, 549–67.

Weller, G.R. 1981. *Politics and the Police: The Case of the Royal Canadian Mounted Police.* Paper presented to the annual conference of the Political Studies Association. Hull: University of Hull.

West Mercia 2001. *The Communities of West Mercia.* [Online: West Mercia Police]. Available at: www.westmercia.police.uk/aboutus/default.htm [Accessed January 2003].

White, R. 2008. *Crimes Against Nature: Environmental Criminology and Ecological Justice.* Cullompton: Willan Publishing.

Wiebrands, C. 1990. Police manpower reallocation: the Dutch case. *Policing and Society*, 1, 57–76.

Wiles, P. and Costello, A. 2000. *The Road to Nowhere. Research Study 207.* London: Home Office.

Williams, F. 1989. *Social Policy: A Critical Introduction.* Oxford: Polity.

Williams, K. and Johnstone, C. 2000. The politics of the selective gaze: closed circuit television and the policing of public space. *Crime, Law and Social Change*, 34, 183–210.

Williams, K., Johnstone, C. and Goodwin, M. 2000. CCTV surveillance in urban Britain: beyond the rhetoric of crime prevention, in *Landscapes of Defence*, edited by Gold, J. and Revill, G. Harlow: Prentice Hall, 168–87.

Williams, T. 1999. *Private Gypsy Site Provision.* Harlow: Advisory Council for the Education of Romany and other Travellers.

Williscroft, C. (ed.) 2007. *A Lasting Legacy: A 125 History of New Zealand Farming Since the First Frozen Meat Shipment.* Auckland: NZ Rural Press.

Wilson, D. and Sutton, A. 2004. Watched over or over-watched? Open street CCTV in Australia. *The Australian and New Zealand Journal of Criminology*, 372, 211–30.

Wilson, J. and Dalton, E. 2007. *Human Trafficking in Ohio: Markets, Responses, and Considerations.* Santa Monica, California: Rand Corporation.

Wolfe, R. 2006. *A Short History of Sheep in New Zealand.* Auckland: Random House.

Women's Aid 2007. *Home Affairs Select Committee Select Inquiry into Domestic Violence. Evidence from the Women's Aid Federation of England.* [Online: Women's Aid]. Available at: http://www.womensaid.org.uk/domestic-violence-articles.asp?section=00010001002200430001&itemid=1650 [Accessed 1st September 2007].

Women's Institute 2008. *No More Violence Against Women Campaign.* [Online: Women's Institute]. Available at: http://www.thewi.org.uk/standard.aspx?id=13492 [Accessed 30 May 2009].

Wood, D.S. 2002. Explanations of employment turnover among Alaska village public safety officers. *Journal of Criminal Justice*, 30, 197–215.

Wood, D.S. and Trostle, L.C. 1997. The nonenforcement role of police in western Alaska and the eastern Canadian arctic: an analysis of police tasks in remote arctic communities. *Journal of Criminal Justice*, 25, 367–79.

Wood, P. 2005. *Dirt: Filth and Decay in a New World Arcadia.* Auckland: Auckland University Press.

Woods, M. 2004. Politics and protest in the contemporary countryside, in *Geographies of Rural Cultures and Societies*, edited by Holloway, L. and Kneafsey, M. Aldershot: Ashgate, 103–25.

Woods, M. 2005. *Contesting Rurality: Politics in the British Countryside*. Aldershot: Ashgate.

Woods, M. 2006. Political articulation: the modalities of new critical politics of rural citizenship, in *Handbook of Rural Studies*, edited by Cloke, P. *et al.* London: Sage, 457–71.

Woods, M. (ed.) 2008. *New Labour's Countryside: Rural Policy in Britain since 1997.* Bristol: Policy Press.

Woods, M. and Goodwin, M. 2003. Applying the rural: governance and policy in rural areas, in *Country Visions*, edited by Cloke, P. Harlow: Pearson, 245–62.

Wu, J. and Phipps, M. 2002. *An Investigation of the Ivory Market in Taiwan.* Cambridge: TRAFFIC.

Yarwood, R. 2001. Crime and policing in the British countryside: some agendas for contemporary geographical research. *Sociologia Ruralis*, 41, 201–19.

Yarwood, R. 2002. *Countryside Conflicts*. Sheffield: The Geographical Association.

Yarwood, R. 2003. A rural policeman's lot is not a happy one: issues and efforts to police rural space in the United Kingdom, in *Rural Services and Social Exclusion*, edited by Higgs, G. London: Pion, 174–87.

Yarwood, R. 2005. Crime concern and policing in the countryside: evidence from Parish Councils in West Mercia Constabulary, England. *Policing and Society*, 15, 63–82.

Yarwood, R. 2007a. The geographies of policing. *Progress in Human Geography*, 31, 447–65.

Yarwood, R. 2007b. Getting just deserts? Policing, governance and rurality in Western Australia. *Geoforum*, 38.2, 339–352.

Yarwood, R. 2008. Policing policy and policy policing: directions in rural policing under New Labour, in *New Labour's Countryside*, edited by Woods, M. Bristol: Policy Press, 205–20.

Yarwood, R. 2010. An exclusive countryside? Crime concern, social exclusion and community policing in two english villages. *Policing and Society*, 20, 61–79.

Yarwood, R. forthcoming. Neighbourhood Watch, in *The International Encyclopaedia of Housing and Home*, edited by Smith, S. *et al.* Oxford: Elsevier.

Yarwood, R. and Edwards, B. 1995. Voluntary action in rural areas: the case of neighbourhood watch. *Journal of Rural Studies*, 11, 447–59.

Yarwood, R. and Cozens, C. 2004. Constable countryside? Police perspectives on rural Britain, in *Geographies of Rural Cultures and Societies*, edited by Holloway, L. and Kneafsey, M. Aldershot: Ashgate, 145–72.

Yarwood, R. and Gardener G. 2000. Fear of crime, culture and the countryside. *Area*, 32, 403–11.

Young, J. 1988. Risk of crime and fear of crime: a realist critique of survey based assumptions, in *Victims of Crime: A New Deal*, edited by Maguire, M. and Pointing, J. Milton Keynes: Open University Press, 164–76.

Young, J. 1999. *The Exclusive Society.* London: Sage.

Young, J. 2002. Critical criminology in the twenty-first century: critique, irony and the always unfinished, in *Critical Criminology: Issues, Debates, Challenges*, edited by Carrington, K. and Hogg, R. Cullompton: Willan, 251–75.

Young, M. 1993. *In the Sticks: Cultural Identity in a Rural Police Force.* Oxford: Clarendon Press.

Zaubermann, R. 1998a. Gendarmerie et gens du voyage en région parisienne. *Cahiers Internationaux de Sociologie*, 105, 415–38.

Zaubermann, R. 1998b. La répression des infractions routières: le gendarme comme juge, *Sociologie du Travail*, 40, 43–64.

Zedner, L. 2009. *Security.* Abingdon: Routledge.

Zvekic, U. 1998. *Criminal Victimisation in Countries in Transition.* Rome: UNICRI.

Index